Also by Klaus Truemper

Mathematics

The Daring Invention of Logarithm Tables
(English and German edition)

The Construction of Mathematics

Brain Science

Subconscious Blunders
Artificial Intelligence
Wittgenstein and Brain Science
Magic, Error, and Terror

Aviation

Lessons from Piloting for 45 Years

Technical

Logic-based Intelligent Systems
Effective Logic Computation
Matroid Theory

Edited by Ingrid and Klaus Truemper

F. Hülster *Introduction to Wittgenstein's*
Tractatus Logico-Philosophicus
(English and German edition)

F. Hülster *Berlin 1945: Surviving the Collapse*

KEY CONCEPTS
OF MATHEMATICS

AN EASY INTRODUCTION

KLAUS TRUEMPER

Leibniz Company

Softcover published by Leibniz Company
2304 Cliffside Drive
Plano, Texas, 75023
USA

The book is typeset in LaTeX using the Tufte-style book class, which was inspired by the work of Edward R. Tufte and Richard Feynman.

Sources and licensing arrangements for all figures are included in the Notes section.

Library of Congress Cataloging-in-Publication Data
Truemper, Klaus, 1942–

Key Ideas of Mathematics: An Easy Introduction
Includes bibliographical references and subject index.
ISBN 979-8-9924500-0-2
1. Mathematics. 2. Introduction

Contents

1

Introduction

Have you ever complained, "I just can't do math"? Or been baffled by a teacher who explained some mathematical concept while you thought, "Why should this be true?" Or been overwhelmed by a flood of mathematical results and wondered, "Who on earth dreamed all this up?"

If any of this applies to you, or if you are just curious how the astounding field of mathematics came to be, this book is for you.

The fact is: There is nothing simple or self-evident about mathematics. Teachers often ignore this and introduce mathematical concepts as if they were the most natural thing in the world. They then show how the concepts can be applied to problems, and ask students to solve additional cases. Effectively, the teacher acts as if it were straightforward to "get it." But it isn't so for many students, who blame themselves for their failure to understand the material, let alone solve problems. They conclude, "I can't do math," or worse yet, "I hate math."

When one delves into the history of mathematics, a very different picture emerges. Brilliant mathematicians of the past developed concepts in extraordinary efforts that often spanned decades or even centuries. That's completely at odds with the way today's teachers often state the results, as if the mathematical claims were the most natural things to consider. No wonder many students have difficulty grasping mathematical concepts.

In contrast, this book shows how mathematical concepts were developed gradually, and how mathematicians often struggled with difficult aspects and only after much effort achieved clarity.

Reading the book and tracing the struggles of mathematicians of the past, you become familiar with fundamental concepts of math, begin to feel comfortable dealing with them, and are ready to delve into details. One student commented after reading an earlier version of this material, "I found myself drawing deep connections among mathematical ideas that I've struggled to understand throughout my education."

The book covers a number of key areas of mathematics. It starts with the idea of number, proceeds to topics such as calculus and infinity, and concludes with the achievements of modern logic. Doing so, the book builds a bridge spanning the entire development of mathematics.

Mind you, the coverage of mathematics in this book is far from complete. But by reading this easily understood book, you become well-prepared to examine other historical developments of mathematics, understand key ideas, and delve into technical details.[1]

———————————

The chapters to follow don't require any math background beyond everyday knowledge of numbers and the elementary operations of addition, subtraction, multiplication, and division. The subsequent Notes section expands upon the discussion and justifies various statements using at most high-school math.

Let's start.

2

Numbers

We all know the *whole numbers*, which consist of 1, 2, 3, ..., their negative counterparts $-1, -2, -3, ...$, and the 0. We also know the *fractions* such as $\frac{1}{2}$ or $\frac{3}{5}$, where a whole number is divided by a nonzero whole number.

These numbers were developed over tens of thousands of years for commerce and trade. Today's computers still use those numbers, and in some sense, these are the only numbers computers know and are able to process.

But for mathematicians, the idea of whole numbers and fractions, useful as they are, did not suffice. So over the past 4,000 years, they created a variety of additional numbers.

At the core of this creative process are the *decimal numbers*. Examples are 0.34714... or 7.31739..., where the digits may stop at some point or continue indefinitely. How many more numbers were so created? The astounding result is: Many, many more beyond the whole numbers and fractions. Let's try to understand this with an analogy.

Abundance of Decimal Numbers

Consider the oceans of the earth. They contain more than 300 million cubic miles of water.[2] Let this huge volume of water represent the number of decimal numbers. Which portion of that volume corresponds to the whole numbers and fractions?

Until the mid-19th century, this question wasn't posed let alone answered, due to the simplistic notion that there were an infinite number of whole numbers, of fractions, and of decimal numbers, and that one could not say more about these quantities. But then Georg Cantor (1845–1918) started a revolution by showing that there are various types of infinity.

Georg Cantor.[3]

Cantor's results imply that the water representing the whole numbers and fractions, among the vast amount of water representing the decimal numbers, constitutes less than a drop of water! In fact, it is less than one molecule of water. Indeed, it can only be said there is *no* smallest amount of water that would represent the whole numbers and fractions!

Thus, mathematicians have created an incredible number of numbers beyond those needed for commerce and trade. But they have also created an astounding *variety* of numbers. The rest of this chapter covers that development.

To start, we need to introduce standard mathematical terminology. Whole numbers are called *integers*, fractions are called *rational numbers* since they are ratios of whole numbers, and *decimal numbers* are a particular representation of the *real numbers*. From now on, we will use that standard terminology. We begin with the integers.

Integers

Counting began at least 50,000 years ago.[4] We will skip a detailed discussion of the innovative ways in which counts were defined, for example, by *body parts*.[5] Around 10,000 BCE, herder-farmers invented sophisticated pebble counting where, for example, 10 pebbles were replaced by one specially marked pebble.[6]

As commerce and trade developed, so did the representation of numbers, computational processes, and methods for recording results. We defer discussion of that development until Chapter 3, and instead cover here the evolution of numbers using the decimal notation of modern mathematics.

We start out with the positive integers 1, 2, 3, ... They usually are called the *natural numbers*. A seemingly trivial observation was that a natural number may be produced by several different multiplication steps. For example, $2 \cdot 12$ is equal to 24, as is $3 \cdot 8$. This fact has little use in commerce, but the ancient Greeks thought it worthy of investigation.

The first observation was that some natural numbers n are not the result of any multiplication step except for the trivial $1 \cdot n$. Examples are $n = 2, 3$, and 5. Such numbers are now called *primes*.[7]

The second observation was that multiplication just using prime numbers can create any natural number. For example, in the multiplication step $2 \cdot 12 = 24$, the *factor* 12 is the result of the multiplication step $2 \cdot 6 = 12$. Hence, we can write $2 \cdot 2 \cdot 6 = 24$. Continuing in this fashion, we can replace each nonprime factor by the product of two smaller numbers, until we have a final representation where all factors are primes. Here, we get $2 \cdot 2 \cdot 2 \cdot 3 = 24$.

These facts piqued the curiosity of the ancient Greeks. In particular, Euclid (mid 4th–mid 3th century BCE) posed the following two questions.

First, are there an infinite number of primes? Euclid proved that this is so.[8]

Second, is the representation of a natural number by prime factors unique except for trivial reordering of the factors? Euclid also proved this to be the case.[9] This result is now known as the *Fundamental Theorem of Arithmetic*.[10]

The primes occur within the natural numbers in a seemingly irregular pattern. Mathematicians since Euclid have striven for a predictive formula, and in the process have created an astonishing wealth of results. They are brought together in the book *The Music of the Primes*.[11]

Division of an integer quantity by another integer quantity, an important operation for commerce, forced creation of the rational numbers; examples are $\frac{1}{4}$ and $\frac{2}{3}$. We look at these numbers next.

Euclid, by Justus van Gent, ca. 1474.[12]

Rational Numbers

In ancient times, it was believed that all measurements in nature could be expressed with rational numbers, using a suitable unit length for each measurement. That belief postulated a unity of nature and mathematics that simply does not exist. The decisive blow came by a member of the Pythagorean society during the 5th century BCE, possibly Hippasus of Metapontum.[13] He showed that the length of the diagonal of a square with length 1 for each side, which was known[14] to be $\sqrt{2}$, is not a rational number.[15]

We should mention that the ancient Babylonians already knew the length of the diagonal to be $\sqrt{2}$ more than 1,000 years before Pythagoras. Indeed, they computed the value of $\sqrt{2}$ with high precision, as proved by clay tablet YBC 7289, which was created

Clay tablet YBC 7289, ca. 1800–1600 BCE. Size about 8cm each side.[16]

around 1800–1600 BCE. It displays the square and shows on the diagonal for $\sqrt{2}$ a Babylonian number that in decimal notation is 1.41421297 and thus correct for five digits after the decimal point. The Babylonian number below the diagonal approximates $\frac{1}{\sqrt{2}}$.

A number that is not rational is called *irrational*. Thus, $\sqrt{2}$ is an irrational number. We look at the irrational numbers next.

Irrational Numbers

How can we tell rational and irrational numbers apart? For a simple rule, we need to look at the conversion of rational numbers to decimal numbers. The conversion is also called *decimal expansion*. For example, $\frac{1}{4}$ becomes 0.25, and $\frac{3}{4}$ becomes 0.75. The decimal expansion need not have finite length. For example, $\frac{2}{3}$ becomes 0.666... and thus involves an infinite sequence of digits.

In each infinite case, the expansion has a particular structure. That is, from some point on in the decimal expansion, a string is repeated over and over. For example, $\frac{628}{2475}$ has the decimal expansion 0.25373737.... Thus, the expansion begins with 0.25, then repeats the string 37 indefinitely.

The irrational numbers also have infinite decimal representation, but not with the repetitive structure exhibited by the rational numbers. This is the decisive difference.

Real Numbers

The *real numbers* are the collection of rational and irrational numbers. An immediate question is: Can the real numbers be systematically constructed? The rational numbers are easily handled by systematic enumeration of ratios of integers. But how can the irrational numbers be constructed? Naïvely, we could think of creating an irrational number as follows: We write down a finite number of digits, then enter the decimal point, and from then on keep on writing digits indefinitely while avoiding any repetitive pattern. The description is technically correct, but gives no insight how the irrational numbers are distributed between the rational ones. Richard Dedekind (1831–1916) described an ingenious construction of the irrational numbers from the rational ones that provides that insight.[17] The construction eventually became know as the *Dedekind cut*. It can be summarized as follows.

Richard Dedekind.[18]

Sort the rational numbers, then cut the sorted list into two lists. Under a certain condition, an irrational number lies in the cut.[19] When this process is repeated for all cuts, all irrational numbers are produced.

Algebraic Numbers

The Dedekind cut constructs all irrational numbers from the rational ones. But so far we have seen just one explicit example of an irrational number, $\sqrt{2}$. How can we construct more irrational numbers? The example $\sqrt{2}$ gave mathematicians an idea: $\sqrt{2}$ is the solution of the equation $x^2 = 2$, which is the same as $x^2 - 2 = 0$. So they defined equations with a single variable x where the right-hand side is 0, and where the left-hand side is a sum of terms $a_k \cdot x^k$ where $k \geq 0$ and a_k are integers. The left-hand side is called a *polynomial with integer coefficients*.[20] Example polynomials are $3x^2 + 15x - 27$ and $36x^5 - 17x^2 - 19$. A solution of the equation is called a *root* of the polynomial.

A number is *algebraic* if it is the root of some polynomial with integer coefficients. The construction of the algebraic numbers has the nice feature that the integers and the rational numbers[21] are algebraic. But there are also lots of examples where a root is irrational. For example, for any integer $k \geq 2$ and any positive integer n, the polynomial $x^k - n$ produces irrational roots if n is not equal to some integer raised to the kth power.

Given the abundance of irrational numbers that are algebraic, the question immediately arose: Are *all* irrational numbers actually algebraic and thus roots of polynomials with integer coefficients?

Leonhard Euler, by Jakob Emanuel Handmann, 1756.[22]

The answer was long in coming. Leonhard Euler (1707–1783) defined the name *transcendental* for irrational non-algebraic numbers, but evidently did not provide an example.[24]

Joseph Liouville (1809–1882) first proved the existence of transcendental numbers[25] in 1844. So, yes, there are irrational numbers that are not algebraic.

Joseph Liouville.[23]

Imaginary Numbers

The computation of roots of polynomials goes back to ancient times. We will skip discussion of the varied history and mention only that time and again a pesky problem surfaced: A root could not be computed because, for some positive number a, a term of the form $\sqrt{-a}$ occurred. Indeed, there is no real number such that multiplication with itself produces a negative number, and thus it was argued that a number $\sqrt{-a}$ did not exist. Eventually, the definition $i = \sqrt{-1}$ was used to reformulate $\sqrt{-a}$ as $\sqrt{a} \cdot i$. The numbers involving i are called *imaginary*. The motivation for the term was that imaginary numbers, though required for computation, were not considered to be actual numbers. This view persisted up to the time of Euler, who established fundamental results involving imaginary numbers.[26]

Complex Numbers

For real numbers a and b, the term $a + bi$ is defined to be a *complex* number. It can be plotted in the *complex plane*[27] where the x-axis is real and the y-axis is imaginary. In that representation, a complex number can be viewed as an arrow, or technically a *vector*, going from the origin of the plane to the plotted point. Operations with complex numbers then become intuitive steps involving vectors.[28]

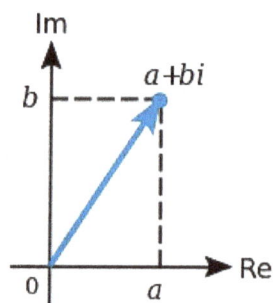

Representation of the complex number $a + bi$ in the complex plane, with real axis Re and imaginary axis Im.[29]

Two Special Numbers: e and π

In subsequent chapters, we repeatedly meet two special numbers: $e = 2.7182\ldots$, called the *Euler number*, and $\pi = 3.1415\ldots$. The number e can be defined in various ways.[30]

For example, $e = \sum_{n=0}^{\infty} \frac{1}{n!}$ where $n! = 1 \cdot 2 \cdot 3 \cdot \ldots \cdot n$.

Jacob Bernoulli (1655–1705) established e while investigating the effect of compounding interest.[32] The number π relates the diameter d of a circle to its circumference c by $c = d \cdot \pi$.

Jacob Bernoulli, by Niklaus Bernoulli (1662-1716).[31]

The numbers e and π are not rational; indeed they are transcendental.[33] Charles Hermite (1822–1901) proved this result for e in 1873, and Ferdinand von Lindemann (1852–1939) for π in 1882 using Hermite's method.

The Creation of Numbers

We have seen a variety of numbers. Where do they come from? Let's begin with the natural numbers. A struggle over tens of thousands of years created the natural numbers[36] and mathematical operations for their use in commerce and trade. That it took such a sustained effort implies that there is nothing obvious about the natural numbers, their representation, or their use. This conclusion in principle does not invalidate metaphysical arguments that, somehow and somewhere outside nature, the numbers and indeed all results of mathematics are stored and ready to be retrieved by the human mind as needed.

Charles Hermite, ca. 1901.[34]

Ferdinand von Lindemann.[35]

In this book, we shall not address these arguments,[37] and only mention the metaphysical claim by Leopold Kronecker (1823–1891) that "God made the integers, all else is the work of man."[38]

The claim was quite appropriate for Kronecker, since he believed in *finitism* of mathematics, where infinite collections are ruled out.[39] But the collection of natural numbers is infinite! Kronecker resolved the conflict between that fact and finitism by appealing to a metaphysical construction.

In the 19th century, it was recognized that the natural numbers, familiar as they were, had not been concisely defined. For discussion of a suitable definition, we need the concept of a *set*, which for our purposes is a collection of *elements*. For example, a set S with three elements a, b, and c is denoted by $S = \{a, b, c\}$. The set of natural numbers, say N, is then $N = \{1, 2, 3, \ldots\}$. We also need the concept of *axioms*, which are statements postulating the existence of mathematical items and relationships among them. Thus, out of thin air, just using the imagination of the human mind, these items and relationships come into existence.

Leopold Kronecker.[40]

Several constructions of N have been proposed.[42] The axioms by Giuseppe Peano (1858-1932), now called *Peano axioms*, are now universally used.[43]

From the set N, we readily define the integers by introducing negation, and the rational numbers by ratios of integers.

Giuseppe Peano.[41]

Next, the Dedekind cut creates the irrational numbers from the rational numbers. The cut is an axiom, as is clearly stated by Dedekind.[44]

Then the algebraic numbers are defined from the integers using roots of polynomials, and the transcendental numbers are all real numbers that are not algebraic.

Finally, we obtain the imaginary and complex numbers from the real numbers by defining $i = \sqrt{-1}$ and, for any real a and b, the number $a \cdot i$ to be imaginary, and the number $a + b \cdot i$ to be complex.

Infinite Proofs

The rules of logic generally do not permit proofs of claims using an infinite number of logic statements. Of course, such a proof couldn't be written down. But one could write down a sequence of arguments terminating with the statement "and so on" and expect that the reader imagines the remaining infinite sequence of arguments. Over centuries, mathematicians constructed proofs of that kind. The claims were couched in more complicated language that avoided the use of "and so on" to disguise the impermissible action.

One such case was the construction of the natural numbers. There are an infinite number of them. How could one possibly define them without using "and so on" or some equivalent statement? It seemed impossible. In the 20th century, Peano solved this problem by including in his construction of the natural numbers the induction axiom. It says the following.[45] If a claim is correct for a base case $n = 0$ and also for arbitrary $n \geq 1$ assuming it is correct for $n - 1$, then it holds for *all* $n \geq 0$.

If you were to remove the induction axiom and its equivalent statements[46] from mathematics, most results could no longer be proved, a devastating effect. The axiom is that important. Yet there is no justification other than the statement, "This axiom makes sense intuitively and is essential for mathematics." You cannot find it in nature, which is one argument supporting the claim that mathematics is a man-made invention and not a discovered part of nature.

Chapter 6 describes profound difficulties connected with the use of "and so on." They were resolved in the 20th century by suitable axioms that go far beyond the induction axiom. Suffice it to say here that on the one hand these axioms turned out to be extremely useful, but on the other hand also permitted proofs of results that in our everyday world would be rejected as crazy. For example, in one instance an axiom called the *axiom of choice* allowed a mathematical doubling of material that is nothing short of miraculous. Chapter 6 has the details.

Summary

The integers together with the rational and irrational numbers constitute the real numbers. The algebraic numbers contain the integers and rational numbers and part of the irrational numbers. The transcendental numbers are the irrational numbers that are not algebraic.

This chapter opened with a comparison of water volumes representing the real numbers versus the natural and rational numbers. The conclusion, based on Cantor's results, implied that the vast volume representing the real numbers is in sharp contrast with the negligible, indeed 0 volume for the natural and rational numbers.

Cantor's results establish an even more astounding conclusion about the algebraic and transcendental numbers. The transcendental numbers, as part of the real numbers, are represented by the entire volume of water, and the algebraic numbers by a volume of 0!

Thus, if we would randomly pick a real number, then with probability 1 it would be transcendental, and it would be a miracle—defined as an event happening with probability 0—that this number would turn out to be algebraic, or more specifically, rational or integer.

Lastly, the induction axiom is a crucial addition to mathematics. It makes proofs possible that otherwise would require the impermissible argument "and so on."

So far, we have ignored issues of notation of numbers and simply have used the decimal system. The next chapter shows that notation for numbers, indeed for all mathematical concepts, is not a trivial aspect of mathematics: Notation can propel mathematical development or stifle, even inhibit, it.

3
Notation

Mathematical notation can propel mathematical development or make it nearly impossible. The reason is the structure of the human brain. Due to evolution, it is eminently equipped to look for patterns. Thus, it performs well when the patterns of notation are indicative of underlying concepts, and it may fail badly when this is not the case. In the jargon of computer science, the brain requires user-friendly notation.

This chapter begins with an example case that is like a laboratory study of the brain's capabilities: Two large groups of mathematicians want to create results in a novel area of research. They are given starting information that, mathematically speaking, is equivalent. For one group, the information is encoded in suitable notation that reflects the underlying concepts. The second group is not so lucky; they must use a notation that does not represent the underlying mathematical structure very well. The laboratory study runs over decades. The mathematicians of the first group consistently produce impressive research results, while those of the second group make little progress.

It's hard to believe that something like this could ever happen. Well, it did—not in a laboratory study, of course, but in the real world. The events were exactly as described, with successful outcomes for one group of mathematicians not just over some decades but over a period of 150 years, and mostly failure for the second group during that time. Here is the story.

Newton versus Leibniz: A Controversy

Calculus is concerned with the following two problems.

First, the problem of *differential calculus*: Given a function of a variable, how rapidly does the function change as the variable changes?

Put differently, what is the slope of the function at a given point? Stated in yet another way, what is the slope of the tangent that touches the function at a given point?

Differential calculus: Tangent straight line touches curved function.[47]

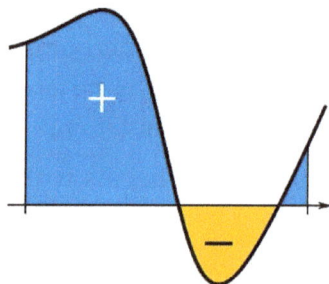

Second, the problem of *integral calculus*: Given a function, what is the size of the area bounded by the function and the horizontal axis?

Here, a certain convention is used. If the function is positive, that is, it lies above the horizontal axis, then the area between the function and the axis is considered to have a positive value. But if the function is negative,

Integral calculus: Area defined by function is total of "+" areas above horizontal axis minus total of "−" areas.[48]

that is, it lies below the horizontal axis, then the area between the function and the axis is considered to have a negative value. The total area is the sum of positive and negative values.

Since ancient times, mathematicians had tried to solve these two problems. They managed to handle special situations, but did not succeed in the general case. In 1666, Isaac Newton (1643–1727) began work on the two problems, and soon found such a method.[49]

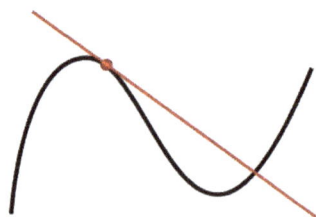

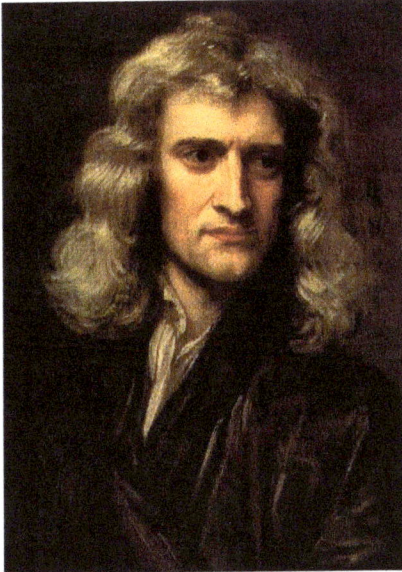

Left: Isaac Newton, by Godfrey Kneller, 1689.[50]
Right: First edition of Newton's Philosophiae Naturalis Principia Mathematica, 1687.[51]

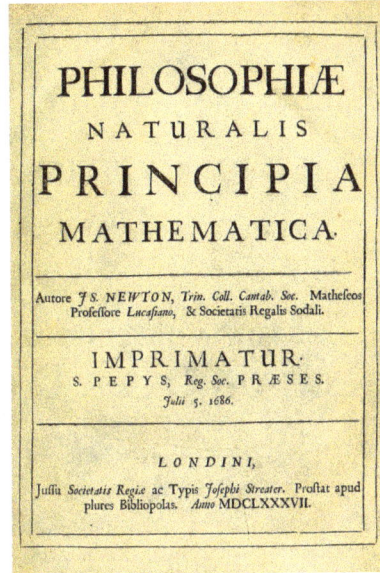

Newton expressed his method using the terms *fluxions* for differential calculus and *fluents* for integral calculus. We shall not go into details here, but mention that these terms were motivated by Newton's concept that changes of quantities represented motion or flow.

Newton used his method extensively in the book *Philosophiae Naturalis Principia Mathematica,* which represented an upheaval in physics. In the three editions of the book appearing in 1687, 1713, and 1726, he described his method[52] for the two calculus problems. Note the delay in publicizing his method: Newton began the work in 1666, but published it for the first time 21 years later!

In a separate development, Gottfried Wilhelm Leibniz (1646–1716) began work on the two problems of calculus in 1674—eight years *after* Newton had begun his effort. For the differential calculus problem, Leibniz defined for a variable x and function $f(x)$ the ratio $\frac{df}{dx}$ where dx is the difference of two x values and df is the difference of the corresponding $f(x)$ values. When dx is chosen infinitesimally small, the ratio constitutes the desired solution. For

ACTA
ERUDITORUM
ANNO MDCLXXXIV
publicata,
ac
SERENISSIMO FRATRUM PARI,
DN. JOHANNI
GEORGIO IV,
Electoratus Saxonici Hæredi,
&
DN. FRIDERICO
AUGUSTO,
Ducibus Saxoniæ &c.&c.&c.
PRINCIPIBUS JUVENTUTIS
dicata.
Cum S.Cæsarea Majestatis & Potentisimi Ele-
ctoris Saxoniæ Privilegiis.

LIPSIÆ,
Prostant apud J. GROSSIUM & J. F. GLETITSCHIUM.
Typis CHRISTOPHORI GüNTHERI.
Anno MDCLXXXIV.

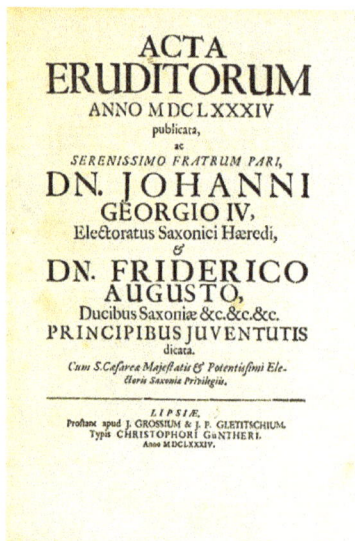

Left: Gottfried Wilhelm von Leibniz, by Christoph Bernhard Francke.[53]
Right: Acta Eruditorum, 1684; contains the first Leibniz publication about calculus.[54]

integral calculus, Leibniz computed the area bounded by the function and the horizontal axis by summing infinitesimal quantities.

Leibniz published his results for the first time in 1684, which means three years *before* publication of the first edition of Newton's Principia. Thus, based on the publication dates, Leibniz had solved the calculus problems first, when in reality Newton had done so earlier.

The story then becomes too complex to be treated here in detail. Initially, Newton and Leibniz acknowledged each other as co-inventors of calculus. But toward the end of the 17th century, misinformation and innuendo by others ignited a bitter feud and ultimately led to Newton and Leibniz accusing each other of plagiarism.

After careful analysis of notes, letters, and publications written by various parties at the time, historians now conclude that Newton first invented calculus, but that Leibniz was not aware of Newton's results and independently invented calculus as well. So, the most fitting conclusion is that Newton and Leibniz are co-inventors.[55]

We draw this conclusion today. But in the 18th century, British mathematicians were convinced that Newton had been grievously wronged by Leibniz. As a result, they rejected Leibniz's approach. In particular, they adhered to Newton's notation, which denotes slope of a function by a dot and integration by a bar.[56]

In contrast, mathematicians on the European continent took Leibniz's side, whose construction uses differences denoted by d and sums denoted by $\int$; the latter symbol represents an elongated "s" of the Latin word "summa." We should emphasize that Leibniz's notation was somewhat different from the modern versions. But the original formulas already reflect the underlying concepts in an intuitive manner.[57]

The superiority of Leibniz's notation quickly became obvious. In a tidal wave of research spanning a 150 year period, continental mathematicians erected a beautiful edifice of ideas and results on Leibniz's foundation. But despite the mounting evidence, British mathematicians continued to adhere to Newton's inferior notation out of loyalty to Newton. As a result, they fell hopelessly behind.

The situation is well expressed by Guillaume de l'Hôpital (1661–1704), who in the first-ever book on calculus[59] writes,

"I must here in justice own (as Mr. Leibnitz himself has done, in Journal des Scavans for August, 1694) that the learned Sir Isaac Newton likewise discovered something like the Calculus Differentialis, as appears by his excellent Principia, published first in the Year 1687 which almost wholly de-

Guillaume de l'Hôpital.[58]

pends on the Use of the said Calculus. But the Method of Mr. Leibnitz's is much more easy and expeditious, on account of the Notation he uses, not to mention the wonderful assistance it affords on many occasions."

The names of the mathematicians building on Leibniz's foundation form an illustrious list. Over the 150 years from 1700 to 1850, it includes, in chronological order, Jacob Bernoulli (1655–1705), Johann Bernoulli (1667–1748), Leonhard Euler (1707–1783), Joseph-Louis Lagrange (1736–1813), Pierre-Simon Laplace (1749–1827), Joseph Fourier (1768–1830), Carl Friedrich Gauss (1777–1855), Augustin-Louis Cauchy (1789–1857), Peter Gustav Lejeune Dirichlet (1805–1859), Karl Weierstrass (1815–1897), and Bernhard Riemann (1826–1866).

It would take too long to even summarize their achievements. But we will review one important problem and its solution. For a moment, let's step back to the time of Newton and Leibniz. A major motivation for the work on differential calculus was the following problem. Given is a function. Where does it attain its maximum and minimum? Differential calculus supplies candidate points for these extrema: They are the points where the slope of the function is 0.[60]

Now consider a more difficult problem: One must find not just maximum and minimum *points* of functions, but must determine *functions themselves* that are best according to some criteria. Here is a problem of that type.

Suppose a straight rail goes from a higher point to a lower point. A piece of metal of some weight is assumed to slide on the rail without friction. Suppose we release the piece at the higher point. Then by the force of gravity the piece will slide to the lower point. We measure the travel time for that movement and wonder: Can we change the shape of the rail so that travel time is reduced? By experimentation, we quickly discover the following. If the rail goes down in a curve with a very steep initial portion, the piece first accelerates quickly and then proceeds with high speed to the lower point. Accordingly, total travel time is reduced.

This fact motivates the following difficult question: What shape must the rail have so that the travel time is as small as possible? The curve achieving the minimum time is called the *brachistochrone curve,*[61] from the ancient Greek βράχιστος χρόνος, which means "shortest time." The question thus is: What is the shape of the brachistochrone curve? This question was first posed by Galileo Galilei (1564–1642) in slightly different form. He thought, mistakenly, that the curve should be an arc of a circle.

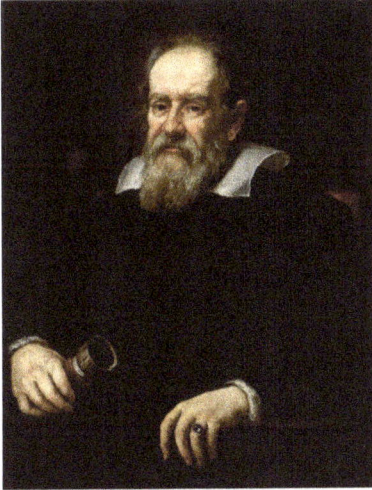

Left: Galileo Galilei, by Justus Sustermans, 1636.[62]
Right: Johann Bernoulli, by Johann Rudolf Huber, ca. 1740.[63]

Johann Bernoulli was the first to determine the shape of the brachistochrone curve. He then posed the problem as a challenge; Newton, Jacob Bernoulli, Leibniz, l'Hôpital, and Ehrenfried Walther von Tschirnhaus (1651–1708) responded with solutions.[64]

Newton had no difficulty solving the problem. He came home from the Mint, where he had the position of Master, saw that there was a letter from Johann Bernoulli, ate supper, read the letter, solved the problem, and wrote down the solution. All this in one evening. The next day he posted the solution letter to Johann Bernoulli.[65] The story proves that Newton had no difficulties working with his notation. It's just that others could not use it as effectively.

The problem of the shape of the brachistochrone curve and its solution motivated investigation of other problems where best functions had to be found. The research area is now called *calculus of variations.* The mathematicians cited earlier for their contributions to calculus as well as many others created a complex body of theory for this area.[66] The entire effort built upon differential and integral calculus, using Leibniz's notation.

Outdoor demonstration: Ring slides down faster on curved brachistochrone rod than on the straight rod.[67]

In the second half of the 20th century, the problems of calculus of variations were recast in a novel way, thus greatly simplifying their solution. The key was the *principle of optimality.* It is most easily explained by an example.

Suppose we travel from New York to Los Angeles by car via the shortest route. If that route goes through, say, St. Louis, then the principle of optimality says that the portion of the route from New York to St. Louis must be, by itself, the shortest route connecting those two cities.[69]

This age-old and seemingly trivial principle was investigated by Richard E. Bellman (1920–1984), who used it

Richard E. Bellman.[68]

to develop a profound methodology called *dynamic programming*[70] for solution of a variety of optimization problems.

Later, that versatile methodology was combined with the results of differential and integral calculus already known to Newton and Leibniz, resulting in the comparatively easy solution of a large portion of the problems of calculus of variations.[71] The dynamic programming approach is still used today. For example, the US space agency NASA employs it to solve a broad variety of control problems arising in space exploration.[72]

The remaining sections of this chapter look at various cases in the history of mathematics where a new notation or small change of an existing concept led to major developments.

Bürgi and Napier: Multiplication Becomes Addition

In ancient times, complicated notation of numbers made multiplication and division quite difficult.[73] But even with the subsequently developed, much simpler notation of decimal numbers, multiplication and division were tedious, and computation of powers and roots was utterly arduous. Mathematicians found relief in two ways. First, by transforming manual computation using exponents—the subject of this section—and second, by using various mechanical devices.[74]

The development of the concept of exponents, trivial as it might seem now, took a long time. For example, in the ancient Greek text *Arithmetica* written by Diophantus in the third century,[75] we have the following confusing notation for powers of unknown quantities,[76] where the top row is the modern notation, and the bottom row displays the corresponding expression of the *Arithmetica*.

x	x^2	x^3	x^4	x^5	x^6
δ	Δ^Y	K^Y	$\Delta^Y\Delta$	ΔK^Y	$K^Y K$

DIOPHANTI
ALEXANDRINI
ARITHMETICORVM
LIBRI SEX.
ET DE NVMERIS MVLTANGVLIS
LIBER VNVS.
Nunc primùm Graecè & Latinè editi, atque absolutissimis Commentariis illustrati.
AVCTORE CLAVDIO GASPARE BACHETO
MEZIRIACO SEBVSIANO.V.C.

LVTETIAE PARISIORVM,
Sumptibus SEBASTIANI CRAMOISY, via
Iacobæa, sub Ciconiis.
M. DC. XXI.
CVM PRIVILEGIO REGIS.

Arithmetica, 1621 edition, translated into Latin from Greek by Claude Gaspard Bachet de Méziriac.[77]

At the middle of the 16th century, the mathematician Michael Stifel (1487–1567) writes the following two rows of numbers:[78]

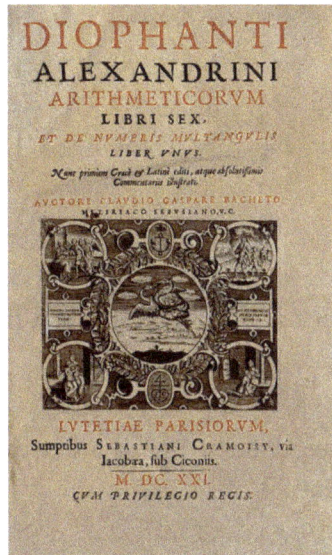

−3	−2	−1	0	1	2	3
$\frac{1}{8}$	$\frac{1}{4}$	$\frac{1}{2}$	1	2	4	8

In the top row, consecutive entries are defined by *adding* 1. Thus, they constitute an *arithmetic progression*. In the bottom row, consecutive entries are produced by *multiplying* by 2. So this is a *geometric progression*.

The numbers are linked as follows. When 2 is raised to the power of a

Michael Stifel.[79]

number of the top row, then the result is displayed by the entry below. Indeed, $2^{-3} = \frac{1}{8}$, $2^{-2} = \frac{1}{4}$, ..., $2^3 = 8$. The "2" in these equations is now called the *base* of the geometric progression. Stifel calls the numbers of the top row *exponents* because they are, well, "exposed." Now comes Stifel's crucial observation:

Suppose we *multiply* two numbers in the bottom row, getting a third number. When we *add* the exponents of those two numbers, we get the exponent of the third number. In modern notation, for any m and n, we have $2^m \cdot 2^n = 2^{m+n}$.

As a result, if we want to multiply two numbers of the bottom row, we simply add their exponents, go to the position where that sum occurs as exponent, and find the result below that exponent.[80]

Stifel's process simplifies multiplication of numbers to addition of exponents. Similarly, division of numbers becomes subtraction of exponents, taking of powers of a number is reduced to multiplication of its exponent by the specified power, and computation of roots of a number is accomplished by division of its exponent by the specified root.[81]

Stifel established these relationships, but could not use them effectively: After all, the simplified computations could only be carried out if the given numbers occurred in the bottom row. But that row was far from containing all possible numbers.

Toward the end of the 16th century, Jost Bürgi (1552–1632), a master craftsman of clocks and precision machinery as well as an accomplished mathematician,[82] saw a way out of the conundrum faced by Stifel.

Bürgi recognized that Stifel's choice of 2 as base in the definition of the geometric progression was just arbitrary.[85] Actually, any positive number other than 1 would do.

Diß buch zeiget künſtlich aen
wie begriffen werden kan
Mathematiſcher inſtrument
Dryangels gehaimnus bhent.
DVRCH WISZENHAIT DISER KVNST
ERLANGT ICH GROSZER HERRN GVNST

Jost Bürgi.[83]

Indeed, if the base was very close to 1, the geometric progression would have very small gaps that could be handled by interpolation. For that reason, Bürgi decided on the base value $B = 1 + \frac{1}{10000} = 1.0001$. He computed the numbers for B^n, $n = 0, 1, 2, \ldots$, 23027, where the case $n = 0$ corresponds to the number 1.0, and $n = 23027$ to 9.99999779, which essentially is 10.0. He streamlined the computations so that they likely required just a few months of manual effort.[86]

No additional numbers needed to be computed beyond the cases $n = 0, 1, 2, \ldots$, 23027 since any positive number of the decimal system is readily scaled by a power of 10 so that it lies between 1.0 and 10.0. Thus, use of the tables only required trivial initial and final scaling by powers of 10.

Mechanized celestial globe, by Jost Bürgi, 1594, Schweizerisches Landesmuseum, Zurich.[84]

Alas, Bürgi did not publicize his results in a timely fashion. In fact, he never published the mathematical reasoning behind his method. Instead, in 1620—about 20 years *after* he had completed the research[87]—he published the above described tables[88] and a manual for their use.[89] The manual explains how the tables are used for multiplication, division, and computation of powers and roots. The discussion includes interpolation of table values.

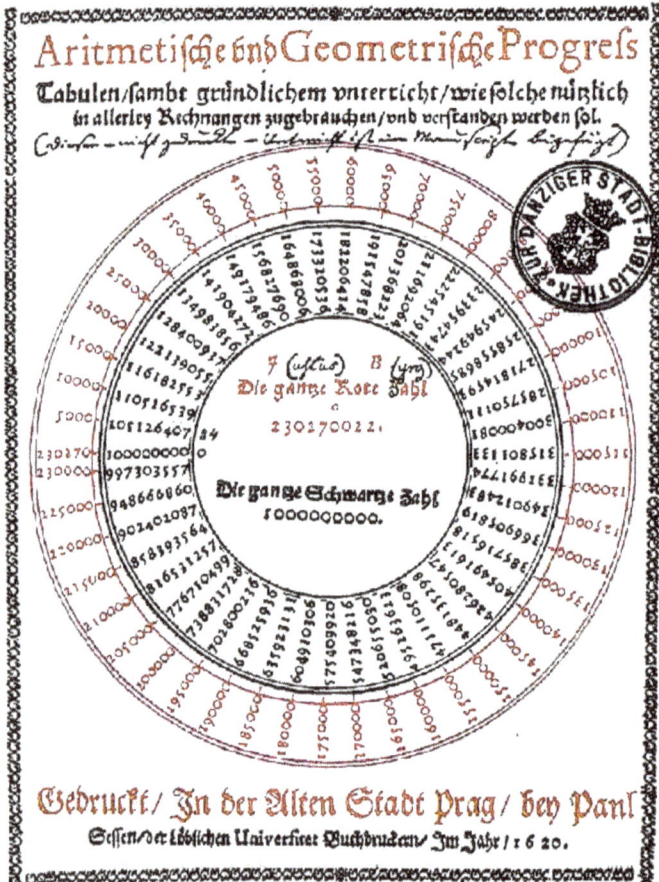

Title page of Jost Bürgi's Table of Logarithms, 1620. The page lists every 500th entry of the table. The numbers are in the inner ring, and the logarithms in the outer ring.[90]

Independently of Bürgi, John Napier (1550–1617)—mathematician, physicist, and astronomer[91]—developed a somewhat different concept of logarithm.[92]

In 1614—six years *before* Bürgi published his manual and tables—Napier presented the results in the book *Mirifici logarithmorum canonis descriptio.* It contained 57 pages explaining the method and 90 pages with logarithm data for the conversion of a number to Napier's logarithm and vice versa.

Left: John Napier.[93]
Right: Napier's Mirifici logarithmorum canonis descriptio, 1614.[94]

The mathematician Henry Briggs (1561–1630) offered[95] to improve the utility of Napier's data by rescaling them to base-10 logarithms. Napier agreed to the transformation. Thus, Briggs carried out the scaling and published the resulting tables. They are sometimes known as *Briggsian logarithms* in his honor.

There have been various discussions whether Bürgi or Napier first developed the idea of logarithm. We shall not weigh in on that controversy. But it is reasonable to conclude that Bürgi and Napier invented the idea of logarithm independently and approximately at the same time; that Bürgi failed to publish his results in timely fashion and did not work out further details; and that Napier investigated logarithms and their use with extraordinary dedication and care.[96]

Descartes: Geometry Becomes Algebra

In 1637, René Descartes (1596–1650)—outstanding philosopher, mathematician, and scientist[97]—described his research methods for interpreting nature in the book *Discours de la méthode*.[98] In an appendix titled *La Géométrie*, he proposed a revolutionary mathematical concept that allowed treatment of geometric problems by methods of algebra.

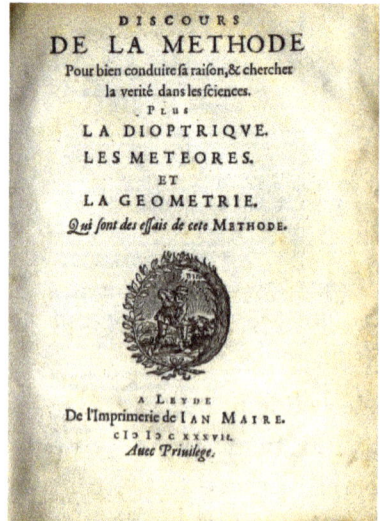

Left: René Descartes, after Frans Hals, 1648.[99]
Right: Descartes's Discours de la méthode, 1637.[100]

Up to Descartes's time, geometry had been a favorite area of mathematical investigation since geometric bodies—for example, point, line, parabola, plane, circle, cylinder, cone, pyramid, sphere—naturally embody a number of relationships that do not require explicit mathematical specification.

For example, "sphere" means a 3-dimensional body with a center point from which all points on the surface have the same distance. It also has a surface area and volume.

At the same time, important relationships connect these objects. For example, the angles formed by the lines of a triangle add up to 180 degrees. Due to these facts, complicated results had been proved in ancient times. We shall not go into details of this development,[101] but mention two outstanding results by the ancient Greek mathematician, physicist, engineer, inventor, and astronomer Archimedes (287(?)–212 BCE). These results plus several others establish

Archimedes, by Domenico Fetti, 1620.[102]

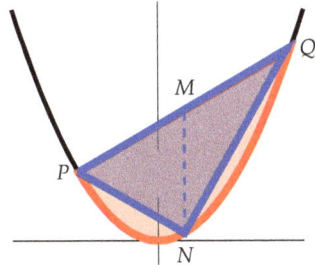

Parabola segment is defined by points P and Q. Triangle is defined by P, Q, and N, where N is vertically below midpoint M of P–Q segment. Area of parabola segment is $\frac{4}{3}$ of area of triangle.[103]

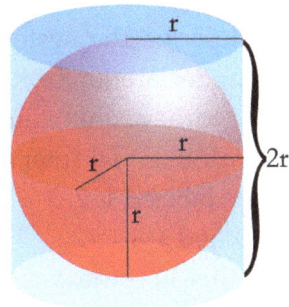

Volume as well as surface area of sphere are equal to $\frac{2}{3}$ that of enclosing cylinder.[104]

Archimedes as the foremost mathematician of antiquity.[105]

The first result concerns a parabola that is cut by a line. The line segment lying within the parabola and a certain point N on the parabola below the line segment define a triangle; see the figure for the definition of N. Archimedes proved that the area within the parabola and below the line is $\frac{4}{3}$ of the area of the triangle.

The second result is the most famous achievement of Archimedes. A sphere is enclosed by a cylinder that is as small as possible. With an amazing sequence of arguments, Archimedes established that the volume of the sphere is $\frac{2}{3}$ of the volume of the cylinder, and that the same ratio applies to the surface areas of the two bodies. Archimedes proved these results with methods that are precursors of the calculus developed 1,900 years later by Newton and Leibniz.

Descartes's *La Géométrie* built a bridge connecting this world of geometry with algebra. To this end, he defined symbols and conventions for constants and variables that are still in use today:

The first letters a, b, c,... of the alphabet denote known quantities, and the final letters ... x, y, z represent unknown quantities.[106]

Powers of quantities are specified by superscript,[107] so for any known or unknown quantity q, the expression $q \cdot q \cdot q \cdot \ldots \cdot q$, where q occurs n times, is denoted by q^n.

In *La Géométrie* Descartes also introduced key ideas that others used later to define the *Cartesian coordinate system*.[108] Indeed, he is often credited with inventing that concept, as evidenced by the name "Cartesian."

We describe the system next. Draw a line, declare an arbitrarily selected point to have the value 0, and then demark in equal intervals additional points labeled 1, 2, 3 ... to the right of the 0 and -1, -2, -3 ... to the left. Call this line the x-axis. Repeat the construction to create a y-axis. Next, place the x- and y-axes perpendicular to each other in the plane, with the x-axis horizontal and the y-axis

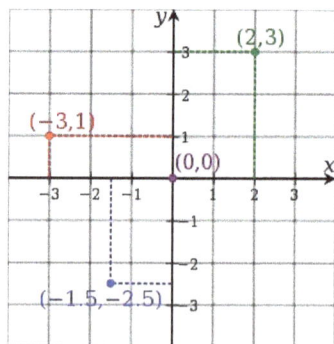

2-dimensional Cartesian coordinate system.[109]

vertical. The axes intersect at their 0 points. The common 0 point is the *origin*.

This configuration is now called the *2-dimensional Cartesian coordinate system*. Any point in the plane is described by its *coordinate values* for x and y as (x, y).

This construction can be extended with another axis, called the z-axis. It is perpendicular to both the x- and y-axes in the obvious way. Now every point of a 3-dimensional world can be defined by a triple (x, y, z) of x-, y-, and z-values in the *3-dimensional Cartesian coordinate system*.

With this machinery, basic concepts of geometry are readily translated into statements of algebra. For example, the points (x, y) of the circle centered at the origin of the plane and with radius r are now defined by the equation $x^2 + y^2 = r^2$. Using such algebraic expressions, results of geometry can be proved much more compactly by algebraic manipulation or by a combination of algebraic manipulation and geometric arguments.

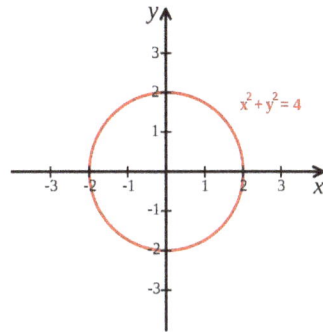

Representation of circle with radius $r = 2$ and centered at the origin.[110]

But Cartesian coordinate systems aren't just a useful device for geometry. Extended to n dimensions by adding axes, they have supported interpretation and manipulation of results in many other branches of mathematics and have become an essential tool for virtually all areas of scientific investigation.

In the next section, we see yet another case where new notation expands the mathematical horizon.

Euler: Formulas Become Functions

Formulas are a core concept of mathematics. When Newton and Leibniz developed calculus, formulas were no longer viewed just as rules of computation, but were analyzed with respect to slope and area.

At that time, Leibniz introduced the term "function" to describe a quantity related to a curve such as slope. Later, the definition of "function" shifted gradually, until in 1755 Euler arrived at a statement that is close to modern usage:

"When certain quantities depend on others in such a way that they undergo a change when the latter change, then the first are called functions of the second. This name has an extremely broad character; it encompasses all the ways in which one quantity can be determined in terms of others."[111]

Euler also introduced the notation "$f(x)$" for functions. With that seemingly simple step, he effectively created a concept of "function" that no longer relied on the specific form of formulas. In the modern definition, this is expressed as follows:

A *function* is a machine that accepts a value of a variable x and outputs the *function value* $f(x)$ for that x value. This concept can be extended to several input variables. For example, the variables may be x and y, and the function is $f(x,y)$.

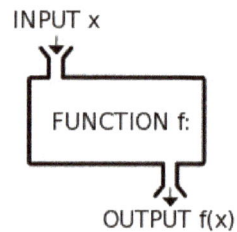

Function as a machine with input x and output $f(x)$.[112]

It often is instructive to display the possible input and output values of the machine by a *graph*.

For example, if x and y are the input variables, then the 3-dimensional Cartesian coordinate system with x-, y-, and z-axis can accommodate all input/output cases as points (x,y,z) using $z = f(x,y)$. A listing of the points of the graph is then just another way to specify the function.

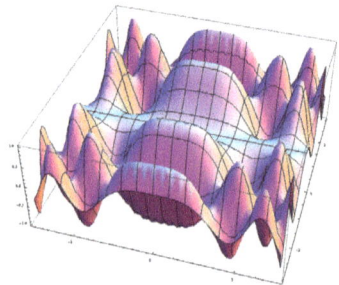

3-dimensional graph of a function $z = f(x,y)$. The x- and y-axes are shown at an angle, and the z-axis is vertical.[113]

Euler's function concept and the $f(x)$ notation shifted the focus from specific rules or formulas to a general view of functions that can be defined, classified, or characterized according to some features.

For example, a function is *continuous* if, informally speaking, it does not jump.[114] Hence, mathematicians began to study continuous functions, defined variations of that property, and looked for conditions that guarantee these versions of continuity to be present.

As a second example, a function $f(x)$ is *differentiable* if for each value of x, the slope of $f(x)$ can be computed. That idea led to conditions that guarantee a function to be differentiable.

The above definition of function also allows for functions for which formulas cannot be specified. On the surface, this may seem like a contradiction: If there is a function, then we must also be able to compute its values, and for that we must have a formula or rule, right? This argument is quite wrong.

In 1936, Alan Turing (1912–1954) proved that there is a function $f(x)$ whose values cannot be calculated by any computer we can imagine. It is called the *halting function*.[115]

The input x of the function is a natural number that represents a computer program.[117] Then $f(x) = 1$ if the program stops after a finite number of steps, and $f(x) = 0$ if the program runs forever. Turing's result implies that there is no formula for the halting function.

Alan Turing, aged 16.[116]

Are there other functions that cannot be evaluated by computer programs? To get some insight into this question, let's take all functions $f(x)$ that for any natural number x have the value $f(x) = 0$ or 1.

Suppose for each such function $f(x)$ we would like to write a computer program that accepts a natural number x as input and then computes $f(x)$. For some functions, this is easy—for example, for the function $f(x)$ that always outputs the value 0. On the other hand, by Turing's result we will be unable to do so for the halting function. Thus we know that we will not be able to write a program for *all* functions. Hence, we modify the goal and write programs for as many of the functions as possible. So imagine that we sit down, pick one of the functions, and try to write a computer program for it. Sometimes we will succeed, sometimes we will fail. Regardless, we then go on to the next function. We stop when we have processed all functions.

For how many functions will we have produced a computer program? For a measure, we take one of the functions at random and ask, "What are the odds that we have written a computer program for this case?" The astonishing answer is that it would be a miracle if we had a program for that function! Mathematically speaking, the probability of having a program for the selected function is 0. An even stronger conclusion applies when we consider functions with slightly different input. That is, the input now is a real number instead of a natural number. The output is 0 or 1 as before. The number of such functions is infinitely larger than that for the functions with natural numbers input and 0 or 1 as output![118]

Evidently, the human mind has created mathematical machinery that is far beyond the need of the practical world of commerce and trade. We saw this first in Chapter 2, where the number of the practically used rational numbers turned out to be an infinitesimal fraction of the number of real numbers.

Euler's work on functions started a virtual flood of mathematical results that looked at functions in various ways.[119] The results justify the claim that the concept of function is one of the most important concepts of mathematics.

Riemann and Lebesgue: Two Ways to Integrate

This section shows how a seemingly trivial shift of viewpoint didn't just expand the power of an already-known mathematical concept, but spawned an entirely new research area. The known concept was integration of functions, and the shift was from vertical slices to horizontal slices of a region.

That shift of thinking led to *measure theory* and the foundation for *probability theory*. The key players in this development were Bernhard Riemann (1826–1866) and Henri Lebesgue (1875–1941).

Left: Georg Friedrich Bernhard Riemann.[120]
Right: Henri Lebesgue.[121]

In 1854, Riemann placed the integration method introduced by Newton and Leibniz on a solid mathematical foundation. Intuitively speaking, Riemann slices the area between the function and x-axis into vertical strips, computes the area for each strip, and then adds up these areas.[122] The process is now known as *Riemann integration*. The method handles many practically important functions, but also fails for a number of instances of interest—for example, when the function jumps so often that any interval of the variable, no matter how small, contains jumps.[123]

In 1902, Lebesgue modified Riemann's process in a seemingly trivial way:

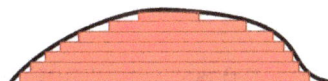

Top: Riemann integration.
Bottom: Lebesgue integration.[124]

He slices the area between the function and the *x*-axis into *horizontal* strips instead of vertical ones.[125]

Lebesgue provided an intuitive description of the process:[126]

"I have to pay a certain sum, which I have collected in my pocket. I take the bills and coins out of my pocket and give them to the creditor in the order I find them until I have reached the total sum. This is the Riemann integral.

"But I can proceed differently. After I have taken all the money out of my pocket I order the bills and coins according to identical values and then I pay the several heaps one after the other to the creditor. This is my integral."

For a demonstration of Lebesgue integration, consider the function $f(x)$ that is equal to 1 if x is rational and equal to 0 if x is irrational. This function cannot be plotted, since the rational and irrational numbers are too closely intertwined. For the same reason, the Riemann integral of $f(x)$ cannot be computed.

In contrast,[127] the Lebesgue integral can be computed, with the conclusion that the area between the function and the *x*-axis is 0, an amazing conclusion![128] When this result is restated in terms of probabilities, we get a proof of the claim of Chapter 2: The probability that a randomly chosen real number turns out to be a rational number is 0.

The main difficulty of Lebesgue integration is measurement of the length of the horizontal strips. Research on that aspect led to an entire theory of measurement. Thus, the seemingly simple change of slicing areas horizontally instead of vertically resulted in a major advance in mathematics.

Summary

We have seen how notation can have a decisive influence on the development of mathematics: Differences of notation hindered or promoted mathematical advances, and seemingly small changes in notation or concepts led to new and profound mathematical results.

———————

The next chapter shows how seemingly reasonable, intuitive thinking about mathematics can be completely wrong. One cause for such failure is the mistaken belief that mathematics is part of nature, that we have a fundamental grasp of nature, and that therefore we have a priori a solid understanding of mathematics that cannot mislead us.

4

Infinity

The human brain has a ready concept for the word "infinity." For example, when the brain hears "an infinite row of trees," it generates a picture of trees neatly arranged in a row extending to the horizon. In the distance, the trees become smaller and smaller until they turn into dots and cannot be discerned anymore. The brain supplements that image with the thought that the trees go on and on beyond the horizon.

The term "infinitesimal" also triggers convincing pictures. For example, upon hearing the definition "A point is a line shrunk to infinitesimal length," the brain envisions a line segment, making it shorter and shorter until it eventually becomes a point.

In the context of poetry or religious texts, the vivid representation of "infinite" and "infinitesimal" causes at most minor problems. In that world, inconsistent information is tolerated, sometimes even purposely introduced, and the brain feels free to create pleasing images.

In mathematics, the brain readily offers the appealing but deceptive belief that "infinity" and "infinitesimal" are actually some sort of numbers.

Sure, these are not regular numbers, but with a bit of care we should be able to link them with the rational or real numbers already on hand. This can indeed be done with suitable restrictions.

But that interpretation can also produce paradoxes and other undesirable mathematical complications.

In this chapter, we look at instances of such interpretations and their ultimate resolution by entirely different viewpoints. The story begins in the first half of the 17th century, a time of scientific discovery and mathematical innovation, but also of oppression of novel ideas.

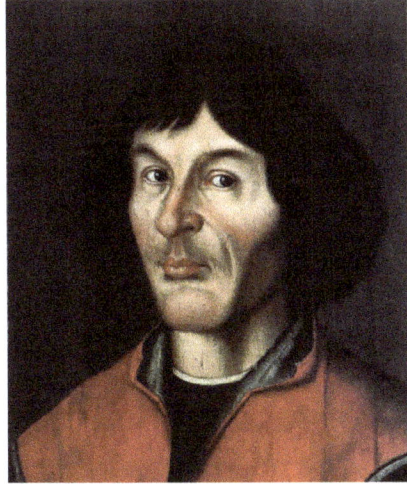

Left: Galileo demonstrates the telescope to Leonardo Donato,[129] the 90th Doge of Venice, and the Venetian Senate, by H. J. Detouche, 1754, detail.[130]
Right: Nicolaus Copernicus.[131]

Galileo, the extraordinary astronomer, physicist, mathematician, and engineer, had learned about the Dutch invention of the telescope. Without having seen the device, he constructed one. He was the first person to direct the telescope toward the sky, making astonishing discoveries such as the moons of Jupiter, sun spots, and the mountains and craters of the earth's moon. As he scanned the night sky, he became more and more convinced that the ancient model placing the earth at the center was incorrect and should be replaced by the sun-centered model of Nicolaus Copernicus (1473–1543).

At the time, the Jesuits of the Society of Jesus were empowered by the Catholic Church to decide validity of any claim about heaven and earth, be it religious doctrine, philosophical or scientific results, or even theorems of mathematics. The Jesuits roundly rejected Galileo's conclusion, maintaining that the earth was the center of the universe.[132] Threatened with torture,[133] Galileo recanted his views of a universe with the sun at the center and lived under house arrest near Florence until his death in 1642.

In this tumultuous world of controversy and oppression,[134] Bonaventura Cavalieri (1598–1647) and Evangelista Torricelli (1608–1647) laid the foundation for modern calculus.

Cavalieri: Method of Indivisibles

Cavalieri[135] viewed a planar area to be composed of an indefinite number of parallel indivisible line segments, and a body to be composed of an indefinite number of parallel indivisible planar areas.

Cavalieri used these concepts to compute areas of surfaces and volumes of bodies by comparing them with known areas and surfaces.

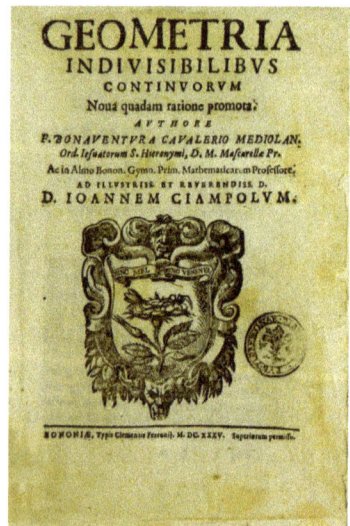

Left: Bonaventura Cavalieri.[136]
Right: Cavalieri's Geometria indivisibilibus continuorum nova quadam ratione promota, 1635.[137]

A modern implementation of his method of indivisibles is called *Cavalieri's principle.*[138] It can be stated as follows:

- Suppose two regions of the plane lie between two parallel lines. Further suppose that every line parallel to these lines intersects the two regions in such a way that the line segments falling within the two regions have equal length. Then the two regions have the same area.

- Suppose two bodies in 3-dimensional space lie between two parallel planes. Further suppose that every plane that is parallel to these planes intersects the two bodies such that the intersections have the same area. Then the two bodies have the same volume.

In 1635, Cavalieri published his method in the book *Geometria indivisibilibus continuorum nova quadam ratione promota*[139] (Geometry, developed by a new method through the indivisibles of the continua). At 711 pages, the exposition is very detailed. He communicated extensively about his work with Galileo, who was very much impressed and concluded, "Few, if any, since Archimedes, have delved as far and as deep into the science of geometry."[140]

Galileo's comment referred to the very first methods for integral calculus developed by Archimedes.[141] Expressed in the terminology of Cavalieri, one could say that for each problem Archimedes used a particular collection of indivisibles. Building upon these groundbreaking ideas, Cavalieri created a unifying method using indivisible line segments and areas.

Torricelli: Extension of Indivisibles

Cavalieri's method crucially depends on parallel indivisible lines that have no width and parallel indivisible planes that have no thickness.

In a powerful extension, Torricelli,[142] a student of Cavalieri, considered an indivisible line to be a very thin rectangle, and an indivisible plane to be a very thin slice. This approach allowed the comparison of indivisible lines or indivisible planes that are not parallel. Torricelli's method is best explained by an example.

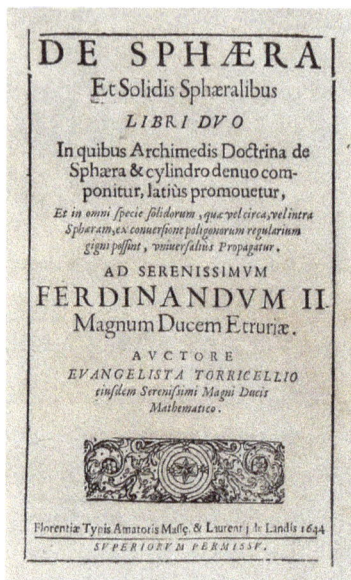

DE SPHÆRA
Et Solidis Sphæralibus
LIBRI DVO
In quibus Archimedis Doctrina de
Sphæra & cylindro denuo com-
ponitur, latiùs promouetur,
Et in omni specie solidorum, quæ vel circa, vel intra
Sphæram, ex conuersione poligonorum regularium
gigni possunt, vniuersalius Propagatur.
AD SERENISSIMVM
FERDINANDVM II.
Magnum Ducem Etruriæ.
AVCTORE
EVANGELISTA TORRICELLIO
eiusdem Serenissimi Magni Ducis
Mathematico.

Florentiæ Typis Amatoris Masse, & Laurentij & Landis 1644.
SVPERIORVM PERMISSV.

Left: Evangelista Torricelli, by Lorenzo Lippi, ca. 1647.[143]
Right: Torricelli's Opera geometrica, 1644.[144]

Consider the drawing of an upright rectangle and a slanted parallelogram. They have the same baseline and the same height.

The vertical strip A demarked by the two dashed lines within the rectangle is an indivisible line segment of the rectangle. Corresponding to strip A is the slanted strip B demarked by two dashed lines within the parallelogram.

The two strips share the same portion of the baseline and have the same area. But strip A is shorter than strip B, and correspondingly wider.

Now suppose that a plane region is composed of strips of type A, and that a second plane region has strips of type B in one-to-one correspondence with the strips of type A.

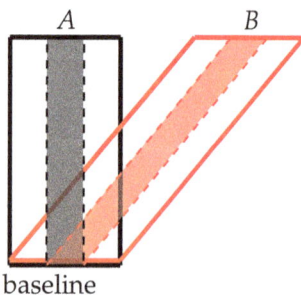

Strip A of rectangle and strip B of parallelogram have same area but different width and length.[145]

Further suppose we know the area of the first region, and that we have a way to compute the ratios between the lengths of the strips of type *A* versus those of type *B*.

Then we can compute the area of the second region using the fact of one-to-one correspondence of the two types of strips and the ratios of strip lengths.

Torricelli published his method in the book *Opera Geometrica* (Geometric Works) in 1644. In just 150 pages, it provided a wealth of results that significantly enlarged the foundation for integral calculus.

Among the bodies investigated by Torricelli was a horn-shaped surface now called *Torricelli's trumpet* or *Gabriel's horn*. It is defined via the portion of the hyperbolic curve $f(x) = \frac{1}{x}$ where $x \geq 1$. The trumpet results when that portion is rotated around the *x*-axis.

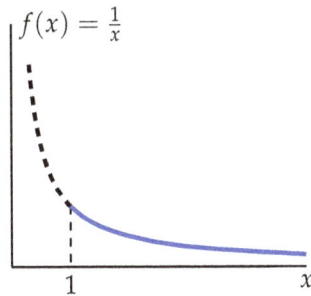

Hyperbola. When the solid portion of the curve is rotated around the *x*-axis, the surface so generated is Torricelli's trumpet.[146]

Torricelli determined that the volume enclosed by the trumpet is finite, but that the surface area is infinite.

It seemed paradoxical that an infinite surface could enclose a finite volume.[148] Since the volume of the trumpet is finite, conceptually it can be filled with a finite amount of paint.

Torricelli's trumpet.[147]

But a finite amount of paint cannot coat the infinite surface of the trumpet, a seemingly contradictory conclusion.

The apparent contradiction is due to an erroneous application of arguments from physics to mathematics: A coat of real-world paint, which has a certain thickness and thus is a 3-dimensional body, is compared with the 2-dimensional surface of the mathematical trumpet.[149]

The work of Cavalieri and Torricelli was strongly supported by Galileo and other mathematicians. But there was a fierce negative reaction by the Jesuits of the Catholic Church.

They continuously monitored research efforts in the sciences and mathematics, looking for consistency with the supposedly eternal truths found in the Holy Scriptures or stated by Aristotle. Indeed, the Jesuits determined that the idea of indivisibles directly contradicted those eternal truths, just as Galileo's claim of a sun-centered universe was heresy.

In a long fight against the use of indivisibles, the Jesuits were ultimately victorious. Coupled with the suppression of Galileo's claim of a sun-centered universe, that victory had a devastating impact on progress in the sciences and mathematics in Italy.[150]

The next step in the understanding of infinity and infinitesimals was taken in the second half of the 17th century in England, far from the oppressive reach of the Jesuits.

Wallis: Infinity and Infinitesimal are Numbers

For the computation of the surface area of a geometrical object, Cavalieri and Torricelli used an indefinite number of indivisible line segments. The term "indefinite" meant that there could be many line segments but that the actual quantity was ignored. The term "indivisible" meant that the width of the line segments could be very small and the exact size needed not be of concern.

The vagueness was well advised:

If the quantity was infinite, then the width couldn't be positive, since such width would make the surface of the given geometrical object infinite. But if the width was 0, the surface area would be 0 as well.

If the quantity was finite, then the width couldn't be very small since otherwise the computed surface area would become very small, too.

Left: John Wallis, by Godfrey Kneller.[151]
Right: Detail of Proposition 3 of De sectionibus conicis (On Conic Sections) by John Wallis, 1655: three triangular figures with height A and baseline B, each with an infinite number of horizontal strips.[152]

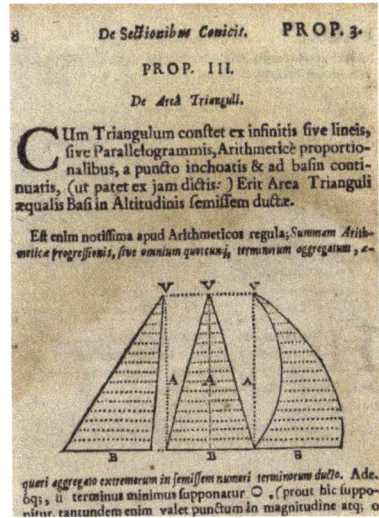

In England, John Wallis (1616–1703) saw this vagueness in the method of indivisibles of Cavalieri and Torricelli and boldly eliminated it by explicit use of infinity and infinitesimal as numbers.[153]

To this end, he defined an infinite number of strips for the computation of surface areas. Each strip had infinitesimal width. He did not try to justify this idea mathematically, but simply argued that his method made intuitive sense and produced correct results.

For computations, he introduced the symbol ∞ for infinity and denoted infinitesimal by $\frac{1}{\infty}$. He used these symbols as if they represented any other number, taking into account that ∞ was very large and $\frac{1}{\infty}$ very small. For example, $\infty \cdot \frac{1}{\infty} = 1$ since he canceled the occurrences of ∞ in the numerator and denominator.[154]

Wallis verified mathematical claims just by checking instances, a generally unacceptable method. But in doing so, he produced impressive results in trigonometry, calculus, geometry, and the analysis of infinite sums.[155] They include a way to compute π with arbitrary precision with the formula $\frac{\pi}{2} = \frac{2}{1} \cdot \frac{2}{3} \cdot \frac{4}{3} \cdot \frac{4}{5} \cdot \frac{6}{5} \cdot \frac{6}{7} \cdot \frac{8}{7} \cdot \frac{8}{9} \cdots$, now known as the *Wallis product*.[156]

Newton and Leibniz: Infinitesimals Become Zero

Cavalieri, Torricelli, and Wallis developed calculus methods for particular classes of problems. The goal of Newton and Leibniz was different: They wanted to solve the general case where a function[157] is given and one desires the slope of the function or the area formed by it and the x-axis.

Newton and Leibniz succeeded admirably. With his calculus method, Newton produced a stunning mathematical foundation of the physical world that stood unchallenged until the time of Einstein. Researchers using Leibniz's approach to calculus developed a huge part of mathematics.[158]

Though Newton's and Leibniz's methods proved to be extraordinarily effective, the arguments supporting them were not fully satisfactory. Perfection would come later, when infinity and infinitesimals were better understood. Let's look at the details of the difficulty.

Consider a function $f(x)$. How rapidly does the function change as x changes? Newton and Leibniz answer the question by considering a small change[159] d of x. The new point is $x + d$, and its function value is $f(x + d)$.

Next, they consider how much the function changes as x becomes $x + d$, that is, $f(x + d) - f(x)$.

Computation of slope.[160]

For example, let $f(x) = x^2$. Then $f(x + d) = (x + d)^2 = x^2 + 2xd + d^2$ and the difference of $f(x + d)$ and $f(x)$ is $f(x + d) - f(x) = x^2 + 2xd + d^2 - x^2 = 2xd + d^2$. At this point, the methods of Newton and Leibniz diverge.

Newton converts the difference $2xd + d^2$ into a *rate of change*, called by him the *fluxion* of $f(x)$, as follows: He first divides the difference

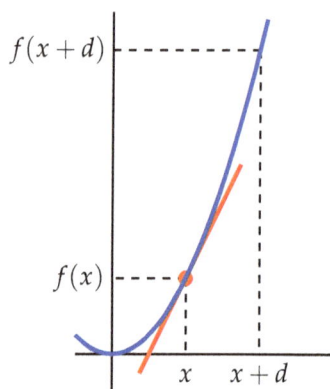

$2xd + d^2$ by d to get the rate of change. This is allowed since d is not 0. The resulting ratio is $\frac{2xd+d^2}{d} = 2x + d$.

Now comes the crucial step. He sets d equal to 0, thus reducing the formula derived for the ratio, $2x + d$, to $2x$. This is the desired fluxion of $f(x)$. To summarize, Newton begins by dividing $f(x + d) - f(x)$ by the nonzero d, but then he evaluates the result by setting $d = 0$! Clearly, the process contains an inherent conflict.

In contrast, Leibniz[161] relies on a hierarchy of infinitesimally small quantities. For the evaluation of the difference $2xd + d^2$, d is considered infinitesimally small, and $d^2 = d \cdot d$ is infinitesimally smaller than d and thus infinite-infinitesimally small.

Based on that conclusion, the term d^2 is declared to be 0, so the difference becomes $f(x + d) - f(x) = 2xd$. Dividing both sides of the equation by d, the slope $\frac{f(x+d)-f(x)}{d} = 2x$ results. Note that the latter approach, while different from Newton's, relies on the same idea that very small quantities can sometimes be set to 0.

In the next section, an entirely different view of calculus eliminates these steps. Why didn't Newton or Leibniz discover that solution?

A reasonable answer is that mathematics was still considered part of nature. Indeed, Newton discussed the dilemma of the supposed division by 0 and appealed to an interpretation of motion for resolution of the difficulty.[162]

Leibniz's work was motivated by the computation of tangents for geometrical figures displayed in the Cartesian coordinate system.[163] He approximated the curves of the figures by piecewise linear segments, then shrunk the segments to infinitesimal size.

From that geometrical interpretation, the idea of several types of infinitely small distances emerged, and it seemed reasonable that infinitely small could be nonzero while infinite-infinitely small was 0.

But contrary to these beliefs, nature simply does not provide a clue how mathematics should deal with infinitesimals or supply a hint how to avoid that concept entirely.

Bolzano, Cauchy, Weierstrass: Continuity

Nature remained a compelling model for calculus and infinitesimals up to the beginning of the 19th century. But then a new view emerged where concepts didn't rely on nature-based arguments, but on unambiguous mathematical terms. Key contributors to this development were Bernard Bolzano (1781–1848), Augustin-Louis Cauchy (1789–1857), and Karl Weierstrass (1815–1897).

They realized that justification of the steps of calculus would be possible only when convergence of sequences as well as smoothness of functions had been precisely defined. We discuss their solution in three steps. First, we cover the definition of convergence of infinite sequences.

Step 1: Convergence of Infinite Sequences

Consider the sequence S of fractions $\frac{1}{k}$, $k = 1, 2, 3, \ldots$, that is, $\frac{1}{1}, \frac{1}{2}, \frac{1}{3}, \ldots$. These numbers get ever closer to 0. We express this by saying that the sequence S *converges* to 0, or that 0 is the *limit* of S. The intuitive sense of convergence is captured by the following definition: A sequence S converges to a limit L if its terms eventually get ever closer to that limit.[166]

In the second step, we use the limit concept to characterize so-called continuity of functions.

Bernard Bolzano.[164]

Augustin-Louis Cauchy. Lithography by Zéphirin Belliard after a painting by Jean Roller.[165]

Step 2: Continuity of Functions

Functions that contain no jumps are called *continuous*. Specifically, a function $f(x)$ is *continuous at a point* c if it doesn't jump at the point $x = c$. It is *continuous* if it is continuous at all points.

There are several equivalent definitions for continuity of functions at a given point c. The simplest definition says: A function $f(x)$ is continuous at a point c if for any sequence of x values, say x_1, x_2, x_3,... that converges to c, the associated sequence $f(x_1)$, $f(x_2)$, $f(x_3)$,... converges to $f(c)$.[167]

In the third and final step, we use the continuity concept to extend functions.

Karl Weierstrass.[168]

Step 3: Continuous Extension of Functions

Consider the following situation. We have a formula for a function $f(x)$ that defines the function values for all possible values x except for *exactly one* instance, say when $x = c$. At that point, the formula cannot be applied for some reason.

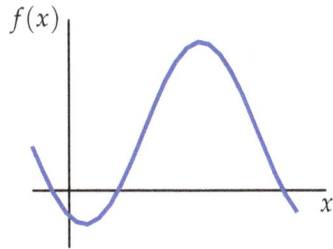

Continuous function.[169]

We want to eliminate this gap in the definition of $f(x)$. We also say that we *extend* the definition of $f(x)$ to the point c.

How should we select that extension? A sensible approach is as follows: We try to find a value L for $f(c)$ so that the function is continuous at c.

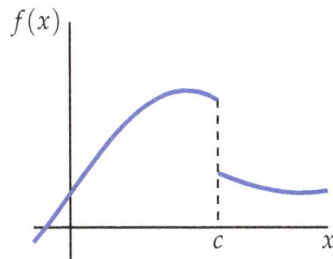

Function with jump at c.[170]

In terms of the continuity definition of Step 2, the selected value L must then observe the following: Whenever a sequence of x values converges to c, then the corresponding function values converge to L.

An abbreviated statement for this latter condition is[171] "the limit of $f(x)$ as x converges to c is L," denoted by $\lim_{x \to c} f(x) = L$.

Thus, the extension process can be written as $f(c) = \lim_{x \to c} f(x)$. We also say that $f(c)$ is the *continuous extension* of the function at c.

The above three steps may now seem simple and almost self-evident. But they were not obvious when Bolzano, Cauchy, and Weierstrass worked to achieve clarity about infinitesimals.

Function $f(x)$ augmented by value L for $x = c$ so that the extended function is continuous at that point.[172]

On the other hand, once these ideas were available, a flood of new insights created a huge area of mathematics now called *Analysis*.[173] It concerns the study of limits, infinite sequences and sums, continuity, differential and integral calculus, calculus of variations, measure theory, and so-called analytic functions, which are defined by certain sums.

We shall not attempt a survey of that area of mathematics. But we do show how the ideas were used to place Newton's and Leibniz's method for the computation of function slope on a solid foundation. We use the example function $f(x) = x^2$ of the preceding section for the demonstration.

Recall that the slope of the function at the point x was estimated via two function values $f(x)$ and $f(x + d)$ as $\frac{f(x+d)-f(x)}{d} = 2x + d$.

Intuitively speaking, we want to eliminate the term d on the right-hand side. Newton did this by setting d directly to 0, while Leibniz eliminated a d^2 term of the intermediate computations. The objection to these steps was, and still is, that such manipulation has no reasonable mathematical justification.

The resolution of the problem relies on the concept of continuous extensions of functions. Define a function $g(d)$ to be the slope

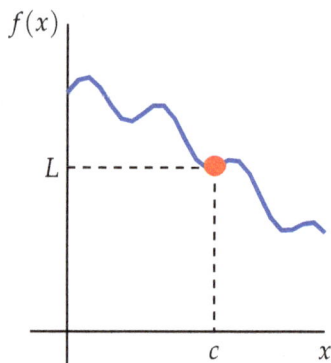

estimate for all nonzero values of d; that is, $g(d) = \frac{f(x+d)-f(x)}{d} = 2x + d$. We emphasize that the definition leaves open the value of $g(d)$ at $d = 0$.

Now comes the crucial break with the approaches of Newton and Leibniz. Instead of eliminating d from the formula $2x + d$ by setting d itself or an earlier encountered d^2 to 0, we obtain the as yet undefined $g(0)$ by the extension process.

That is, we look for a value L for the planned extension $g(0) = L$ so that $g(d)$ is continuous at the point $d = 0$. Now for any sequence $d_1, d_2, d_3, \ldots$ that converges to 0, the sequence $g(d_1), g(d_2), g(d_3), \ldots$, which is $2x + d_1, 2x + d_2, 2x + d_3, \ldots$, converges to $2x + 0 = 2x$. Hence, the continuous extension of $g(d)$ at $d = 0$ is $g(0) = L = 2x$.

The above process was done for an arbitrary point x. Hence it is valid for any x value, and we have $2x$ as the slope function of x^2. In modern notation, we have for $f(x) = x^2$ the slope function, or derivative, $\frac{df}{dx} = 2x$.

The same arguments can be used in the general case of slope computation and thus constitute the long-sought remedy for Newton's and Leibniz's setting certain infinitely small quantities to 0.[174]

The remedy also helped identify functions where the slope cannot be computed. In fact, Weierstrass determined a continuous function, now called the *Weierstrass function*, where the slope cannot be computed for any point.[175]

In a similar fashion, limits also replaced indivisibles and eventually led to the Riemann and Lebesgue integration covered in Chapter 3.

Next we turn to Cantor's fundamental work about infinity.

Cantor: Creation and Classification of Infinities

For more than 2,000 years, mathematicians viewed the term "infinity" with suspicion.

Given the common belief that mathematics was part of nature, how could the concept be explained? Clearly, infinity wasn't a number.

So what did it constitute? And how did it manifest itself in mathematical objects?

For some settings, there seemed to be clear relationships between infinities.

For example, it appeared obvious that a line consists of an infinite number of points, a plane of an infinite number of lines, and 3-dimensional space of an infinite number of planes. Therefore, a line had to have far fewer points than a plane, which in turn had to have far fewer points than 3-dimensional space.

As a second example, the rational numbers are ratios of integers, so clearly there had to be many more rational numbers than integers. Indeed, for any two rational numbers r and s, no matter how close, other rational numbers, for example $\frac{r+s}{2}$, lie between them; on the other hand, consecutive integers do not have this property.

As a third example, the algebraic numbers, defined as the roots of polynomials with integer coefficients—see Chapter 2 for details—include all rational as well as many irrational numbers. So clearly, there had to be many more algebraic numbers than rational numbers.

But there were no precise mathematical concepts that on the one hand captured the intuitive idea of infinity and on the other hand permitted proof of the above conclusions.[176]

There was a sure way to avoid these troubling thoughts about infinity: Simply demand that the concept of infinity couldn't be used in mathematics. This requirement is now known as *finitism*.[177]

Kronecker believed that finitism should be the underpinning of mathematics. Certainly, by its very definition, finitism was guaranteed to eliminate the conundrums about infinity.

But we know now that it also would have imposed a limitation on mathematical creativity akin to limiting the speed of man-made transportation machinery to that of a pedestrian. That restriction would have converted the automobile to a useless device and limited human flight to balloons controlled by a tether.

Even the ancient mathematicians did not believe that finitism was the correct response to the troubling idea of infinity, and by the 17th century, mathematicians accepted infinity as a valid concept used in various, mostly cautious ways.

Good examples of this development are the results of Archimedes, Cavalieri, and Torricelli. The bold approach by Wallis, who simply declared infinity to be a number, is a notable exception. But despite frequent use of infinity, no mathematically sound concept emerged.

There were philosophical ideas about infinity, for example the concept of *potential infinity* and *actual infinity*.[178]

Potential infinity was exhibited by the natural numbers since they could be enumerated by 1, 2, 3, ... without end. *Actual infinity* concerned the case where the natural numbers were collected in a set, which then could be manipulated. Clearly, that set had infinite size, and this motivated the term "actual infinity."

But these and similarly vague definitions didn't lead to a better understanding of the mathematical use of infinity.

In the 19th century, Cantor bursts into this world of vague infinity concepts with a revolutionary approach. He first solves the following basic problem: How can the concept of size or *cardinality* of a finite set, which simply is the number of elements in the set, be extended to infinite sets?

His solution doesn't rely on direct counting. Instead, it *compares* two infinite sets and results in a relative claim about cardinality. The set N of natural numbers is the fundamental yardstick for this process,[179] since it has the smallest cardinality among all infinite sets under a reasonable assumption.[180]

Cantor denotes the cardinality of the yardstick set N by the Hebrew letter aleph with subscript 0, written $\aleph_0$. Any set T having cardinality $\aleph_0$ is said to be *countable*, due to the fact that the elements of T can be lined up and enumerated just as the natural numbers can be listed as 1, 2, 3, ...

For the cardinality measurement of other infinite sets, the following result is useful. Let R and S be countable sets. Derive a set T from R by, informally speaking, replacing each element of R by a copy of all elements of S. The resulting set T turns out to be countable.[181] Let's call this process *substitution* of S into R to create T.

With substitution, it is easily proved that the sets of integers, of rational numbers, indeed of algebraic numbers, have the cardinality of N and thus are countable.[182]

Another simple proof establishes that the sets of algebraic points contained in a line, plane, 3-dimensional space, indeed any n-dimensional space with finite n, are all countable.[183]

Thus, the line contains just as many algebraic points as any n-dimensional space with finite n. At the time, it was an astonishing upset of conventional wisdom!

Next, Cantor shows that there are more real numbers than algebraic numbers, and thus more real numbers than natural or rational numbers. The key element of the ingenious proof is now called Cantor's *diagonal argument*.[184] It links the real numbers with all possible subsets of the set of natural numbers.[185]

Motivated by this relationship, Cantor denotes the cardinality of the set of real numbers by $2^{\aleph_0}$. How much larger is $2^{\aleph_0}$ relative to $\aleph_0$, the cardinality of the set of natural numbers?

Chapter 2 includes a probabilistic statement that compares the abundance of real numbers, measured by $2^{\aleph_0}$, with the sparsity of the rational numbers, captured by $\aleph_0$: If we choose a real number between 0 and 1 at random, then the probability that it is a rational number is 0. The fact that both the set of rational numbers and the set of algebraic numbers have the same cardinality $\aleph_0$, allows us to expand the claim: The probability that a randomly selected real number is algebraic, is 0 as well.[186]

We include one more result of Cantor. Using a mind-boggling construction of so-called *ordinal numbers*,[187] he defines a set of minimum cardinality that is not countable. He denotes the cardinality of that set by $\aleph_1$.

He now has two uncountable sets that seem closely related: The set of real numbers, with cardinality $2^{\aleph_0}$, and the smallest uncountable set, with cardinality $\aleph_1$. He conjectures that the two cardinalities are the same, so $2^{\aleph_0} = \aleph_1$. If true, the real numbers would constitute a smallest uncountable set.

This conjecture is known as the *continuum hypothesis*. For the rest of his life, Cantor tried to prove this hypothesis, to no avail. Indeed, in Chapter 6 it is shown that no such proof is possible.

Seen against the background of the prior, puny concepts of infinity, Cantor's constructions and claims were an extraordinary expansion of mathematics.

Some mathematicians, in particular those believing in finitism, were aghast: Kronecker went so far as to call Cantor a "scientific charlatan," a "renegade," and a "corrupter of youth."[188]

Looking back, the criticism stemmed from a philosophical misunderstanding of the connection between mathematics and the world: The fact that the concept of infinity is not part of nature does not imply that mathematics should not employ that concept. Cantor simply introduced new axioms about infinity and derived their consequences.

Cantor paid a steep price for publishing his results: Kronecker not only undermined Cantor's relationships with other mathematicians, but also made sure that Cantor could not advance to a coveted professorship in Berlin, Germany. The damage Kronecker inflicted on Cantor is evident from Cantor's letters to Gösta Mittag-Leffler (1846–1927), a friend who

Gösta Mittag-Leffler.[189]

strongly supported him.[190] But eventually Cantor's astounding contributions were fully recognized, as we shall see in Chapter 6.

Summary

The concepts of infinity and infinitesimal stemmed from the realization that the natural numbers go on and on, and that rational numbers can become smaller and smaller.

Over centuries, mathematicians tried to capture and treat these two concepts in various ways. In the process, they produced significant results. But lacking was a firm mathematical foundation that fully justified the conclusions.

Construction of that foundation had to wait till the 19th century, when limits and several types of infinity were invented and became powerful mathematical tools for clarifying the elusive ideas of infinity and infinitesimal.

The struggle for clarity in mathematics is also evident in the attempts to solve seemingly simple but actually very difficult problems that were first posed in antiquity and then resisted solution for more than 2,000 years. The next chapter covers six such problems. Amazingly, all of them were completely resolved in the 19th century.

5
Six Problems of Antiquity

As the ancient Greek mathematicians began to explore the landscape of mathematics created by their imagination, they identified a number of fundamental problems. They proceeded to solve many of them. But some cases, though seemingly solvable, turned out to be difficult and resisted solution.

In this chapter, we look at six of these difficult problems, to be defined in detail shortly: Constructing regular polygons, finding roots of polynomials, trisecting angles, doubling the cube, squaring the circle, and redundancy of Euclid's axiom for parallel lines.

It wasn't just the ancient Greeks who couldn't solve these six problems. Over the next 2,000 years, many mathematicians attempted solutions, got partial results, but could not find complete answers.

Finally, in the 19th century, all six problems were solved in rapid succession. What made this possible? A simple answer would be: Over those 2,000 years, more and more concepts and ideas were developed, and finally there was enough insight to solve the six problems.

But there is a more incisive answer. Up to the 18th century, mathematics was considered part of the world, and mathematical methods agreed with physical concepts. But then mathematics began to stand on its own.

Examples of that development are Euler's invention of the function concept[191] and his declaration that the imaginary numbers, which up to that time were considered to be just figments of the imagination, were numbers as much as the real numbers.[192]

The gap between mathematics and the world grew dramatically in the 19th century. For example, number theory was started by Euclid with the concept of prime numbers and, up to the 19th century, was largely concerned with properties of numbers.

But during the 19th century, profound relationships were established that linked numbers with other, often new, mathematical structures. Cantor's construction of transfinite cardinal and ordinal numbers is an example of that development.[193]

Another shift away from nature occurred in geometry, where the firmly held belief in a world shaped according to Euclid's geometry was shattered by a flood of newly created geometries.

These new and often radical ideas made solution of the six long-standing problems possible.

Definition of the Six Problems

In the real world, one can take a straightedge or ruler, draw a straight line, and then use markings of the straightedge to define distances on the line.

Next, one can take a compass and draw a circle using one of the demarked distances as radius. Alternately, if the compass has angle markings, one can draw circles using those markings.

The ancient Greeks created an abstract version of this process for mathematical investigation. They supposed that the straightedge had *no* markings of distance, and that the compass had *no* device for angle measurement.

Since there were no distance markings on the straightedge, the Greeks also defined an initial line segment of arbitrary length and declared it to have distance equal to 1. Except for that single distance, no other tool for distance measurement was provided. What could be drawn with these elementary tools?

By the way, there is a minor technical point. The ancient Greeks assumed that the compass collapses as soon as it is lifted off the plane. This seems to rule out transfer of distances using the compass. But Euclid showed[194] a

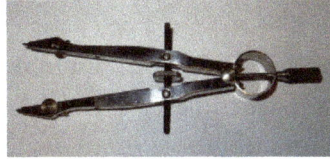

Non-collapsing compass.[195]

construction whereby distances effectively can be transferred by a collapsing compass. Hence, we may assume that the compass does not collapse when lifted off the plane.

From now on it is assumed that all constructions are done with a straightedge without distance markings, a non-collapsing compass without angle markings, and an initial length of 1 defined on a line.

We are ready to state the six problems.

Constructing regular polygons: A regular polygon with $n \geq 3$ edges is obtained from a circle by subdividing the circumference of the circle into n arcs of equal length, then replacing each arc by a straight edge.

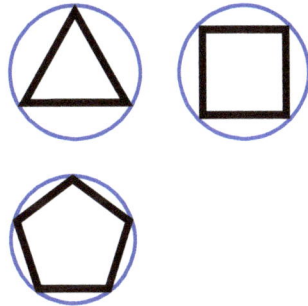

The smallest regular polygons are the equilateral triangle, the square, and the pentagon. The problem demands construction of all regular polygons.

Equilateral triangle, square, and pentagon: the regular polygons with 3, 4, and 5 sides.[196]

Finding roots of polynomials: Chapter 2 discusses the number $\sqrt{2}$ as a solution of the equation $x^2 - 2 = 0$. The left-hand side of the equation is an example of a *quadratic polynomial* with integer coefficients. The general form is $a_2x^2 + a_1x + a_0$, where a_0, a_1, and a_2 are integers. A solution of the equation $a_2x^2 + a_1x + a_0 = 0$ is a *root* of the polynomial $a_2x^2 + a_1x + a_0$. Adding terms, again with integer coefficients, we get the *cubic polynomial* $a_3x^3 + a_2x^2 + a_1x + a_0$, the *quartic polynomial* $a_4x^4 + a_3x^3 + a_2x^2 + a_1x + a_0$, the *quintic polynomial* $a_5x^5 + a_4x^4 + a_3x^3 + a_2x^2 + a_1x + a_0$, and so on. The problem demands the construction of formulas that determine the roots of all such polynomials while using just the four basic arithmetic operations and the taking of the nth root, for any n.

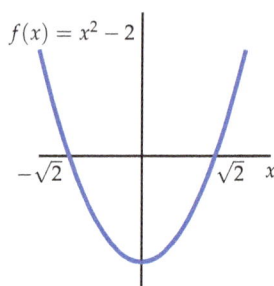

Roots $-\sqrt{2}$ and $\sqrt{2}$ of polynomial $f(x) = x^2 - 2$.[197]

Trisecting angles: Given is an angle defined by two crossing lines. The angle is to be subdivided into three equal angles.

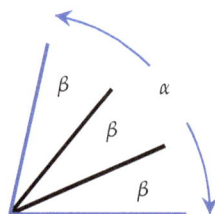

Trisection of angle α into three angles β.[198]

Doubling the cube: Given is a cube. A second cube is to be constructed that has twice the volume of the first one.

Squaring the circle: Given is a circle. A square is to be constructed having the same area as the circle.

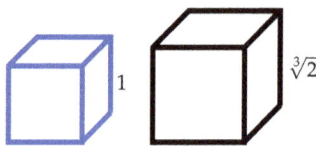

Cube with side length 1 and volume 1, and cube with side length $\sqrt[3]{2}$ and volume 2.[199]

Redundancy of Euclid's axiom for parallel lines: Euclid introduced in the book *Elements* five axioms as the foundation of geometry. When the first four postulates are assumed, the fifth axiom is equivalent to the following:[201]

Given a line and a point not on that line, there is at most one line going through the point that does not intersect, and thus is parallel to, the given line.

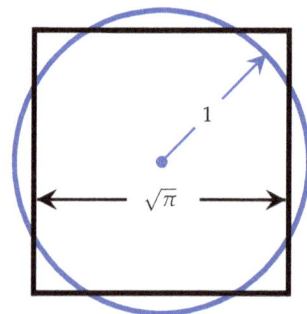

Circle and square with area equal to π.[200]

It seemed utterly obvious that the fifth postulate was implied by the first four. The problem demands that this intuitive idea of the redundancy of the fifth postulate be proved.

Let's see how these six problems were solved.

Given line L and point P: Line M is parallel to L, while line N is not.[202]

Construction of Regular Polygons

The ancient Greeks constructed the triangle, square, and pentagon as well as the 15-sided pentadecagon and the trivial extensions where the number of edges is doubled.[205]

These results provoked the conjecture that maybe all regular polygons could be constructed. But all additional cases resisted solution.

That was the state of knowledge when Carl Friedrich Gauss (1777–1855) considered the problem, at age 19. Gauss arguably became the most eminent mathematician since antiquity.

His subsequent work covered a broad range of areas, including algebra, analysis, astronomy, electrostatics, geometry, geodesy, geophysics, mechanics,

Carl Friedrich Gauss, by Christian Albrecht Jensen, 1840.[203]

Gauss's whimsical signature at age 17.[204]

matrix theory, number theory, optics, and statistics.[206] We get a glimpse of the genius of his work in this chapter.

Gauss had the bold idea to completely disregard the geometric results known at the time and to focus just on five elementary operations on distances that the ancient Greeks had already carried

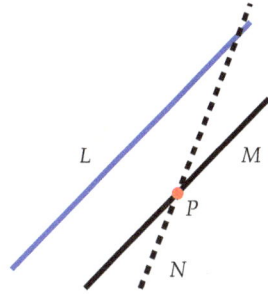

out with geometric constructions: the four basic arithmetic operations of addition, subtraction, multiplication, and division; and the taking of square root.[207]

He also knew that construction of any regular polygon with p corners was equivalent to finding the p roots of the polynomial $x^p - 1$. Indeed, for odd p,[210] the polynomial has one real root, which is $x = 1$, and $p - 1$ complex roots. The p roots are evenly distributed on the unit circle in the complex plane.[211]

When successive roots are connected by line segments, the regular p-sided polygon results. For the case of $p = 17$, two drawings show the 17 roots and the derivation of the regular 17-sided polygon.

The latter drawing is enlarged so that the difference between circle and polygon becomes apparent.

Gauss describes how he came upon the construction of the regular 17-sided polygon as follows:

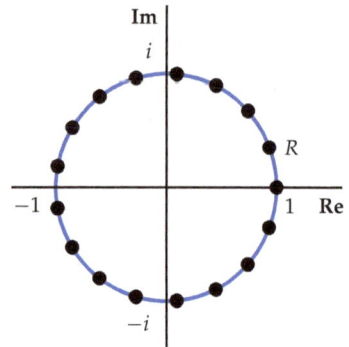

Roots of $x^{17} - 1$ plotted on the unit circle of the complex plane, with real axis Re and imaginary axis Im. Due to the symmetry, derivation of the root $R = \cos\left(\frac{2\pi}{17}\right) + \sin\left(\frac{2\pi}{17}\right)i$ suffices to construct the heptadecagon.[208]

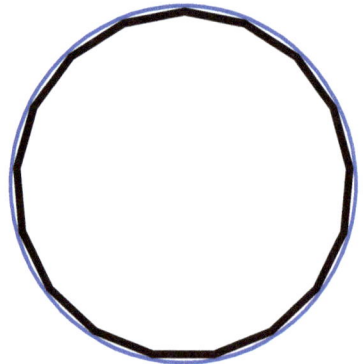

Heptadecagon: the regular polygon with 17 sides.[209]

"The history of this discovery [of the construction of the 17-sided polygon] has up to the present nowhere been publicly alluded to by me; I can give it very exactly, however. The day was March 29, 1796, and chance had absolutely nothing to do with it. Before this, indeed during the winter of 1796 (my first semester in Göttingen), I had already discovered everything related to the separation of the roots of the equation[212] $\frac{x^p - 1}{x - 1}$ into two groups[213].... After intensive consideration of the relation of all the roots to one another on arithmetical grounds, I succeeded during a holiday in Braunschweig, on the morning of the day alluded to (before I had got out of bed), in viewing this relation in the clearest way, so that I could immediately make special application to the 17-side[d] [polygon] and to the numerical verification."[214]

It was the first new construction of a regular polygon since antiquity. Gauss's radical break with history made it possible: He used the complex roots of a polynomial and replaced geometric steps requiring straightedge and compass with the basic operations of arithmetic and the taking of square roots.

Five years later, Gauss constructed an entire collection of regular polygons. His result uses the *Fermat primes* first investigated by the mathematician and lawyer Pierre de Fermat (1607–1665).

They are of the form $2^{(2^n)} + 1$, for $n \geq 0$. Only five Fermat primes are known: 3, 5, 17, 257, and 65537, corresponding to the cases $n = 0, 1, 2, 3$, and 4. Indeed, so far there is no proof establishing either existence or nonexistence of additional Fermat primes.[215]

Pierre de Fermat.[216]

Here is Gauss's result: A regular polygon is constructible by straightedge and compass if the number of edges is a product of distinct Fermat primes and powers of 2.

He conjectured that the condition was also necessary. Pierre Wantzel (1814–1848) proved that conjecture 36 years later.[217] Thus, the polygons specified via Fermat primes and powers of 2 are precisely the constructible regular polygons.[218] This result is now known as the *Gauss-Wantzel theorem*.

Gauss valued the construction of the heptadecagon so much that he requested it to be carved on his tombstone. The stonemason declared that this was impossible[219], since it essentially would look like a circle. A compromise was found for Gauss's memorial in Braunschweig: It shows a star with 17 points.

Star with 17 points at Gauss Memorial, Braunschweig.[220]

Finding Roots of Polynomials

Since antiquity, mathematicians tried to find formulas for the roots of various polynomials. This turned out to be easy for the quadratic polynomials, as we are taught in high school. The roots for cubic polynomials were much harder to determine, but were eventually found in steps that started with special cases and terminated with the most general solution in the 16th century.[223]

The roots of the quartic polynomials were also found in the 16th century by reduction to the cubic case.[224]

Paolo Ruffini.[221]

But the quintic case, where the polynomial generally is of the form $a_5x^5 + a_4x^4 + a_3x^3 + a_2x^2 + a_1x + a_0$, could not be solved despite considerable effort. Then, in the 19th century several mathematicians published results that the quintic case as well as all higher cases were generally unsolvable. Here is a sketch of the developments.[225]

In 1799, Paolo Ruffini (1765–1822) established the result that the quintic and all higher cases in general could not be solved. The proof was incomplete.

Niels Henrik Abel.[222]

In 1824, Niels Henrik Abel (1802–1829) published that same result. The claim contained a flaw that, in hindsight, is not considered major.

In 1845, Wantzel acknowledged the prior work of Ruffini and Abel while publishing another proof. Today, Abel and Ruffini are credited with the result, now known as the *Abel-Ruffini theorem*.

Unaware of all these developments, Évariste Galois (1811–1832) proved

Évariste Galois, age 15.[226]

the result in 1829, at age 18. It was published posthumously in 1843.

Hidden behind this terse summary is a story of misfortune of Abel and Galois worthy of a Greek tragedy. Mismanagement of the submitted papers and bungling of referees prevented timely publication of the papers and proper recognition of Abel and Galois during their short lives. Abel died at age 26 in poverty, his results largely unrecognized; two days later, a letter arrived offering him a professorship in Berlin. Galois died at age 20 from wounds suffered in a duel.[227]

Abel and Galois investigated algebraic manipulations with a new concept now called *group*. A group consists of a set of elements and two operations that are inverse to each other. For example, the set of integers and the operations of addition and subtraction define a group.[228].

A *field* is a more complex structure. It involves a set of elements and four operations. For example, the rational numbers with addition, subtraction, multiplication, and division with nonzeros form a field.

When the set of elements is finite, the group or field is called finite.[229]

Galois created a theory, now called *Galois theory*, that links certain groups with fields. To honor Galois, the finite fields are called *Galois fields.*[230] For each of them, the number of elements is equal to a power of a prime number.

Taken together, the ideas of groups and Galois theory constitute major steps of the 19th century separating mathematics from the real world.

Trisection of Angles

Trisection of angles demands that an arbitrary angle is subdivided into three equal angles using straightedge and compass. Wantzel compared the operation of trisection with the construction steps of straightedge and compass, using polynomials to represent the steps, and thus showed that the construction steps couldn't possibly trisect all possible angles.[231]

His proof that regular polygon construction is possible only when the number of edges is a product of distinct Fermat primes and powers of 2—see the *Gauss-Wantzel theorem* described above—also implies that trisection generally is not possible.[232] The trisection problem can also be solved using Galois theory.[233]

Once more we have seen how new abstract concepts made solution of a long-standing problem possible, demonstrating again how mathematics of the 19th century moved away from concepts of the world. The same conclusion applies to the solution of the next ancient problem.

Doubling the Cube

For the discussion, it suffices that the given cube has side length $x = 1$, and thus has volume $x^3 = 1$. Doubling that cube requires constructing a cube with side length y such that the volume is $y^3 = 2$. Evidently, the side length must be $y = \sqrt[3]{2}$.

Wantzel proved that this was impossible using polynomials, just as he had argued in the angle trisection problem.[234] The result can also be obtained via Galois theory.[235]

The fifth problem, which requires the squaring of the circle, turned out to be much more difficult.

Squaring the Circle

Commerce and trade of ancient times often involved computation of areas of geometric figures. The simplest case was the square: To obtain the area, one only needed to multiply the side length with itself.

The ancient Greeks posed the following general problem: Given a geometric figure in the plane, derive a square with the same area. The problem was called *squaring* the geometric figure.

A number of cases were easily solved. For brevity we will not cover the geometric steps, but simply show, equivalently, that the area could be computed by the four basic arithmetic operations and the taking of square root, just as Gauss did when he constructed the heptadecagon. Once the area is known, taking the square root gives the side length of the equivalent square.

Case of the *triangle*: A simple formula, known since ancient times, says that the area is one half of the length of any side of the triangle times the height of the triangle for that side. The height can be constructed, and thus the area be computed.

Case of any *polygon*, which is a geometric figure in the plane whose boundary consists of straightline segments: Divide the polygon into triangles, and sum up the areas of the triangles.[237]

Case of figures with a curved boundary: In antiquity, the squaring of any

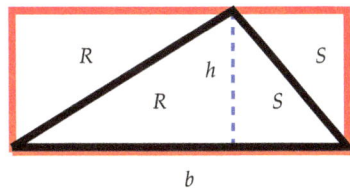

Triangle with baseline b and height h consists of areas R and S. Rectangle has two Rs and two Ss, and total area is $b \cdot h$. Hence triangle area is $\frac{b \cdot h}{2}$.[236]

such figure was considered very difficult.

Hippocrates of Chios (470(?)–410(?) BCE) created the first result for such a geometric figure, now called the *Lune of Hippocrates*. He proved that the lune has the same area as the triangle, and thus can be squared.

The proof mainly consists of a single application of Pythagoras's theorem for right triangles.[238]

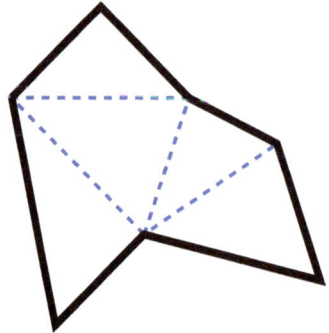

Partition of polygon into triangles.[239]

Archimedes solved a much more difficult problem: He squared any parabola segment. Indeed, he proved that the area of a given parabola segment is $\frac{4}{3}$ of a certain inscribed triangle. Thus, one only needs to construct the triangle and compute $\frac{4}{3}$ of its area.[242]

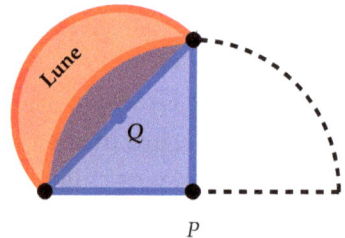

Lune of Hippocrates is defined by quarter circle centered at P and semicircle centered at Q. It has the same area as the triangle.[240]

Archimedes's method was a precursor of integral calculus. As that calculus was gradually developed,[243] the ancient methods for squaring geometric figures became increasingly unimportant, with one exception: the squaring of the circle.

That problem continued to fascinate mathematicians, for the simple reason that squaring seemed obviously possible if one could just find the correct approach. The 19th century

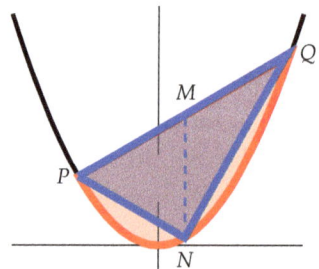

Archimedes: Parabola segment is defined by points P and Q. Triangle is defined by P, Q, and N, where N is vertically below midpoint M of P–Q segment.[241]

brought clarity to this famous problem. For the discussion, we need the concept of real, algebraic, and transcendental numbers introduced in Chapter 2. Recall that the algebraic numbers are the real numbers that are roots of polynomials with integer co-efficients. The transcendental numbers are the real numbers that are not algebraic.

Let's consider squaring of the circle with radius $r = 1$. Its area is $r^2\pi = 1^2\pi = \pi$, and the equivalent square has side length $\sqrt{\pi}$. Squaring the circle then amounts to constructing $\sqrt{\pi}$, or equiva-lently due to multiplication, π.

Wantzel showed that the four basic arithmetic operations and the taking of square root could only produce numbers that were roots of polynomials with integer coefficients. Indeed, the exponents of those polynomials were even.[244] Thus, all numbers produced by that process are algebraic.

This implies that a squaring method producing π with these op-erations could only exist if π was algebraic. The hope that π had this property was dashed in 1882, when Lindemann showed π to be transcendental.[245]

Thus, squaring of the circle was finally proved to be impossible. That result was based on the vastly expanded knowledge about numbers and functions developed in the 19th century.

All problems discussed so far effectively call for geometric con-structions or algebraic formulas. The sixth and last problem is dif-ferent. It demands insight into the role of Euclid's axioms for plane geometry.

Redundancy of Euclid's Parallel Axiom

In his book *Elements*, Euclid introduced five postulates for plane geometry.[246] When the first four postulates are assumed, the fifth is equivalent to the following axiom[247] due to John Playfair (1748–1819):

In a plane, given a line and a point not on it, at most one line parallel to the given line can be drawn through the point.

There are several other axioms that are equivalent to Euclid's fifth.[248] The discussion below also relies on the following:

The sum of the angles in every triangle is 180 degrees.

Mathematicians desire systems of axioms to be minimal; that is, no postulate should be implied by the others.[250] In the case of Euclid's postulates, it seemed utterly obvious that the parallel postulate was implied by the first four. But for 2,000 years, no-

John Playfair, by Henry Raeburn.[249]

body could prove this supposedly self-evident fact.[251] Indeed, when one looks at the drawing demonstrating Euclid's parallel and non-parallel lines, it seems that one can never come up with another system of lines where parallel lines behave differently. The mesmerizing effect of this image was overcome in the 19th century, when a new geometry was proposed that did not use Euclid's fifth axiom.

The history of that new geometry is too complicated to be covered here with reasonable precision. However, the history was compiled[252] with great care at the end of the 19th century when letters and notes of the various contributors were still available. The following summary is based on that material.

At the turn of the 19th century, Gauss was already aware that Euclid's fifth axiom could not be proved and that a different geometry was possible where the sum of angles of any triangle was less than the 180 degrees guaranteed by Euclid's fifth axiom. He hesitated

Ferdinand Karl Schweikart.[253]

to publish this material, fearing a backlash against the new idea. Indeed, he kept quiet about his insight except in letters.

In 1818, he became aware of work by Ferdinand Karl Schweikart (1780–1857), who had concluded that another geometry was possible that he termed *astral geometry*.[254]

In comments about Schweikart's material, Gauss indicated that he was aware of the results.[255] He wrote, "Es ist fast alles mir aus der Seele geschrieben," which was a polite way of saying, "Almost all of it is familiar to me."

In 1824, Franz Adolph Taurinus (1794–1874), a nephew of Schweikart, sent some results about geometry to Gauss. In a long and kind response, Gauss wrote that the results were enjoyable to read, but also pointed out that they were quite incomplete. He added that for more than 30 years he had worked on non-Euclidean geometry, and concluded by requesting—indeed demanding—that Taurinus should view the letter as a private message that he could not publish or reference.[256]

In 1829, Nikolai Lobachevsky (1792–1856) independently developed the new geometry, which eventually came to be called *hyperbolic*. He replaced Euclid's fifth axiom by the following:

Nikolai Lobachevsky, by Lev Kryukov, ca. 1843.[257]

For any given line and any given point not on that line: The plane containing both the line and the point also contains at least two distinct lines through the given point that do not intersect the given line.

There are several ways to display the hyperbolic plane.

Henri Poincaré (1854–1912) proposed a beautiful method, as follows. The hyperbolic plane is represented by a circular disk D with radius 1 in the Euclidean plane.

The hyperbolic points are the points of D that do not lie on the boundary of D, such as point P.

The hyperbolic lines are represented by arcs of circles that lie within D and make a right angle with the boundary of D, such as lines B and C, or by lines that are diameters of D, such as line A.

Lines B and C go through the point P and do not intersect with the line A. According to the definition of "parallel," the lines B and C are thus parallel to line A.[260]

The display preserves angles, and by inspection we can tell that example triangle T has angles summing to less than 180 degrees.

This is entirely different from Euclidean geometry, which can be displayed with straight lines and with exactly one parallel line for any given point outside a given line.

Lobachevsky published his results in Russian; mathematicians outside Russia became aware of his work only years later.

In 1832, János Bolyai (1802–1860) also developed the hyperbolic geometry, again independently.

When Gauss became aware of Bolyai's result, he declared that he had thought of this result decades ago,[262] but also

Jules Henri Poincaré.[258]

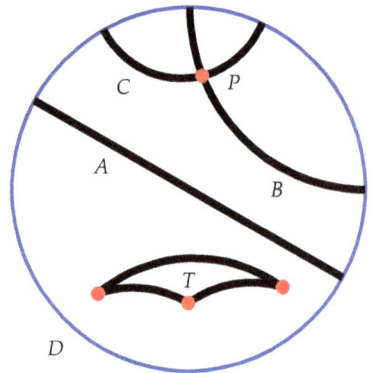

Poincaré's disk D: A, B, and C are lines. T is a triangle.[259]

János Bolyai.[261]

wrote to a friend,[263] "I regard this young geometer Bolyai as a genius of the first order."[264]

Now and then the question is raised how much Gauss had established about the hyperbolic geometry prior to seeing Bolyai's results. Gauss's letters cited above are clear evidence that he fully understood the new geometry. Thus, it is appropriate that Gauss, Lobachevsky, and Bolyai are generally considered to be co-inventors of the hyperbolic geometry.

For a time, it was not clear whether the definition of the hyperbolic geometry contained some inconsistency; that is, whether there was some contradiction inherent in the axioms.

Gauss was troubled by this question, and this likely was one of the major reasons why he did not publish his results.[265]

In 1868, Eugenio Beltrami (1835–1900) resolved this important question[267] by proving that either both Euclidean and hyperbolic geometries were consistent or both were inconsistent.

This result put to rest the lingering suspicion that, somehow, the hyperbolic geometry was inherently flawed. If so, that fate was shared by Euclidean geometry.

The proof that hyperbolic geometry was just as valid as Euclidean geometry eliminated a mental barrier about the nature of geometry.

Eugenio Beltrami.[266]

No longer was there a single geometry that governed all: A change of just one axiom of the Euclidean geometry had resulted in a new, equally acceptable framework.

That insight led to a flood of new geometries, beginning with Riemann in 1853. He expanded the concept of geometry to what is now called *Riemannian geometry*. The hyperbolic geometry is a special case of Riemannian geometry, as is the *elliptic geometry*, where

all lines intersect.[268] The elliptic geometry can be visualized on the surface of a sphere.

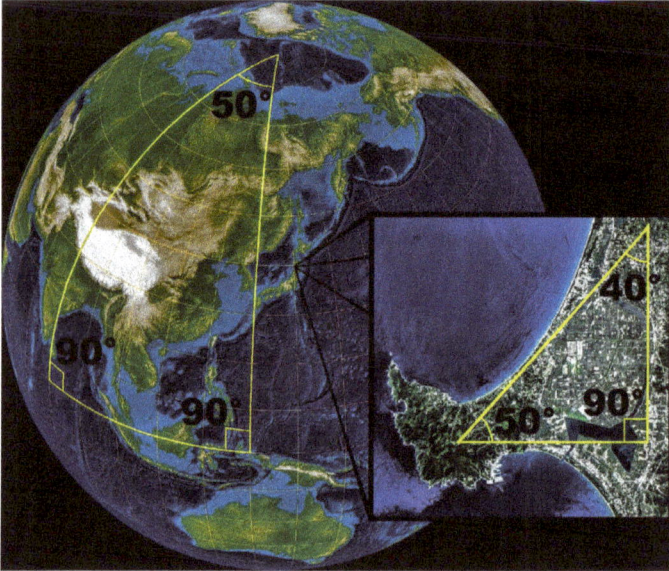

The earth as a spherical model of elliptic geometry.[269]

The great circles of the sphere, which are the circles dividing the sphere into two equal hemispheres,[270] represent the lines.

Any two *antipodal* points, which are points on the sphere opposite to each other, represent together just one point of the geometry.[271]

In that geometry, the sum of angles of a triangle is always greater than 180 degrees. For example, in the large triangle drawn on the earth as an example sphere, the angles sum to 230 degrees. As the triangle becomes smaller and smaller relative to the size of the sphere, the sum of the angles gets ever closer to 180 degrees, as is demonstrated by the triangle in the inset of the picture.

Summary

Six problems of antiquity, open for more than 2,000 years, were all solved in the 19th century. The main reason for that success can be traced to a fundamental change of viewpoint, initiated by Euler in the 18th century: Mathematics is different from nature, does not need nature, and should not be confused with nature.

That thought developed gradually. It met considerable resistance at times, but ultimately prevailed. The results were an array of new mathematical concepts such as groups and fields, and new geometries.

Coupled with new levels of abstraction—for example, the transfinite cardinal and ordinal numbers—this development was nothing short of revolutionary.

―――――――――

Up to this point, we have carried out arguments about mathematical ideas and claims as if there were a natural way to prove results. The next chapter shows that this assumption about mathematical arguments, made since ancient times, has no a priori justification.

Starting in the 19th century, mathematicians made an extensive effort to remedy this misconception. As we shall see, there were major successes, but also humbling failures.

6

Proof

Every day, we face a tsunami of information pouring from newspapers, radio, TV, and the internet. Much of it is just somebody's guess or, worse yet, artful lying.

In response, we reject information unless there is some proof of veracity, such as: A trusted person confirms the information; the data include the results of scientific tests that support the claim; we have knowledge about the subject matter, and the statement is consistent with that information; pictures validate the message; or a trusted website[272] confirms correctness.

Of course, these checks don't really prove validity. They just make it more likely that the information is correct.

Scientists have erected a higher hurdle for proofs of validity; they demand confirmation of claims by repeated trials.

Coupled with Occam's razor[273]—it prefers simpler explanations if there is a choice—scientists have an outstanding track record of weeding out incorrect results and confirming valid ones.

As an aside, the method does have a drawback: Nature may produce very rare events that, even if recorded by chance, will be rejected since they do not recur in subsequent experiments. Thus, science may be creating a picture of a more evenly behaving nature than is actually the case.

Due to experiment-based evaluation, scientists readily modify or even abandon a model or theory when contrary experimental results show up.

For example, in the world of Newton, space extends evenly in all directions, and time flows evenly forever.

For 250 years, that view was considered correct. But in the early 20th century, Albert Einstein (1879–1955) replaced it with the model of *space-time*[276] where gravity bends space and time depends on the relative speed of the observer.

Albert Einstein.[274]

The model is a generalization of the Riemannian geometry of Chapter 5. It has been confirmed by numerous experiments, some of which are ongoing today.

When a scientific theory has been confirmed by numerous experiments conducted over long periods of time, the theory may be *postulated* to be an infallible *law of nature*.

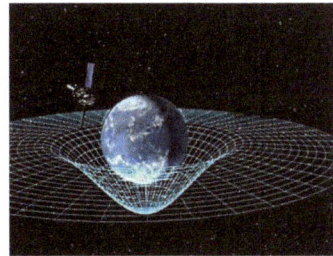

Illustration of a NASA probe orbiting the earth to measure space-time. Note the bending of space by gravity.[275]

For example, when water is heated sufficiently, it starts to boil and becomes steam. This conversion of liquid to steam has been observed for thousands of years, so we have postulated that it is a law of nature.

In mathematics, the bar for proof is even higher: Results cannot be verified just by a number of experiments, but must be shown to be *permanently valid*. Put differently, mathematical results must be correct independently of time and the state of the world. This goal is laudable, but can it be achieved? Two problems stand in the way.

First, any mathematical result in one way or other relies on some initial assumptions and some deductive methods. If the selected assumptions or deduction methods are in dispute, then some mathematicians accept the result while others reject it. So what does "valid" mean in this context?

Second, the precision demanded by mathematicians for assumptions and deductions may increase over time; indeed, it has done so for centuries. Accordingly, a generally accepted result may later—maybe after centuries—be viewed with suspicion. At that time, the result is proved again according to the higher standard, or, if that cannot be achieved, declared to be unproven. However, even if the result is proved again, how can we be sure that the new proof won't be found to be defective at some time in the future?

Early on, mathematicians were aware of these two problems and tried to avoid them with varying degrees of success. By the mid-19th century, enough insight had been gained that mathematicians dared a frontal attack on the problems of precision and validity of results.

That effort lasted roughly 100 years. It produced satisfying insight on many fronts, but also the disturbing conclusion that some results were forever unprovable.

Progress was not always accompanied by pleasant discussions. Indeed, at times there were vehement disagreements that separated mathematicians into camps, with each group claiming perfect insight and correctness. This chapter's main focus is on those 100 years of mathematical developments. They are yet another demonstration of human ingenuity and persistence.

To set the stage, we take a brief look at the proof techniques of antiquity.

Ancient Babylonians: Proof via Drawings

The ancient Babylonians faced a major hurdle when they tried to prove a result. They knew integers and rational numbers, but did not have the modern concept of variable or formula. Nevertheless, they proved interesting results by focusing on geometric problems.

Clay tablet YBC 7289 of 1800–1600 BCE, already discussed in Chapter 2, depicts one such result. The tablet declares a precise approximation to be the length of the diagonal in a square with side length 1. In equivalent decimal notation, it is 1.41421297.

Clay tablet YBC 7289, ca. 1800–1600 BCE.[277]

Another example is clay tablet Plimpton 322, written around 1800 BCE. It lists integer triples a, b, and c that are solutions to the equation $a^2 + b^2 = c^2$.

For a right triangle whose sides have the lengths a, b, and c, with c largest, Pythagoras's theorem[279] establishes this equation. For this reason, triples satisfying the equation are called *Pythagorean*.

Clay tablet Plimpton 322, ca. 1800 BCE. Lists Pythagorean triples.[278]

Rudman argues convincingly[280] that the Babylonians most likely used geometric figures to derive the triples. Based on that analysis, he concludes that the Babylonians knew of Pythagoras's theorem about right triangles at least 1,200 years before Pythagoras.

During the period 350–50 BCE, the Babylonians carried out precise astronomical computations with geometrical methods, a fact discovered by Ossendrijver[281] while analyzing several Babylonian clay tablets. Until that discovery, it was assumed that the astronomers of Babylon solely used arithmetical methods and not geometrical ones.

The key result, shown on the next page in modern notation, is as follows. The velocity of the planet Jupiter is plotted over a time interval of 120 days. Time 0 is the point where Jupiter rises over the horizon, and time 120 is the point of *first station*,[282] where Jupiter seems to stand still before beginning the *retrograde motion*.[283] The total area under the velocity curve then represents the displacement of the planet, measured in degrees, during that time interval.

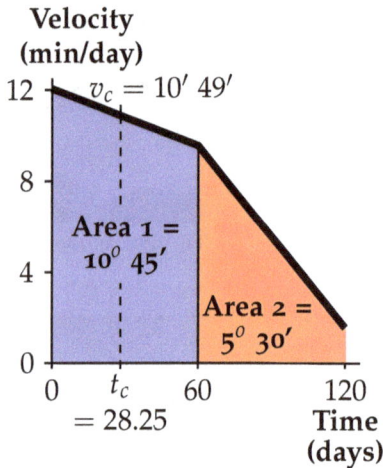

Left: Velocity graph for Jupiter from the time the planet rises on the horizon to the time of first station. The velocity is expressed in minutes per day, where 60 minutes are equal to 1 degree.[284]

Right: Clay tablet BM 34757 of 350–50 BCE contains trapezoid computations of area 1 of the figure.[285]

The figure divides the area under the velocity curve by a vertical line at time 60 into two trapezoids.

The computations for the left trapezoid are recorded on clay tablet BM 34757. During the time period of that trapezoid, Jupiter moves 10^0 45'. During the time of the right trapezoid, Jupiter moves 5^0 30'.

The symbol t_c on the time scale denotes the time when the planet has covered half of the left trapezoid area. The velocity at that time is v_c. The Babylonians computed quite precise values for t_c and v_c, given that the exact solution involves nonterminating fractions of their hexadecimal system of numbers; the value 28.25 of t_c is a rounded decimal. The methods predate related techniques of medieval European scholars by at least 14 centuries.[286]

We move on to the groundbreaking work of the ancient Greeks.

Ancient Greeks: Logic, Axioms, Proofs

Aristotle (384–322 BCE) formalized the logic of deduction in several treatises. Collectively, they are called the *Organon*, which in Greek means "instrument" or "tool." For 2,000 years, this work was considered to be a complete treatment of logic that would never require substantive change or expansion.[287]

In his book *Elements*, Euclid assembled axioms of geometry and, starting with those basic assumptions, proved essentially all results of geometry known at that time.[288]

Archimedes created an array of amazing results that motivated the work of mathematicians for the next 2,000 years.[290] Examples are included in Chapters 3 and 5.

Viewed together, the achievements of just these three mathematicians constitute a profound foundation of mathematics: The results provide a framework for reasoning in proofs, show

Aristotle. Marble, Roman copy after a Greek bronze original by Lysippos from 330 BCE; the alabaster mantle is a modern addition.[289]

how axioms should be formulated and used, and demonstrate that intricate proof techniques can produce astonishing results.

Of course, there were a number of other accomplished mathematicians of antiquity who contributed to fundamental insights. Due to space limitations, we cannot possibly discuss them or their results even in a cursory manner.[291]

Up to the 17th century, the logic of Aristotle stood virtually unchanged. But then Leibniz started a research effort that created many elements of modern logic. Unfortunately, he never published his results. Two mathematicians of the 19th century, George Boole (1815–1864) and Gottlob Frege (1848–1925), independently came up with and expanded upon Leibniz's ideas.

Hence, it is appropriate to consider Leibniz, Boole and Frege to be the pioneers of modern logic. The next three sections cover them in detail.

Leibniz: Anticipation of Modern Logic

Leibniz not only realized that Aristotle's system of logic had considerable shortcomings, but also saw that these defects could not be addressed by a few changes. So he set out to create a new foundation of logic.

The effort was successful, but here and there his construction contained gaps that apparently kept him from publishing the results. As a consequence, his work only became known 150 years later, in the middle of the 19th century. There are two ways to evaluate Leibniz's work on logic.

First, one may examine his results, locate the gaps, fill them, and then compare the now-complete results with modern logic. When this is done,[292] the conclusion emerges that Leibniz anticipated many logic results of the 19th and 20th century.[293]

The second way of evaluating Leibniz's results does not try to fill gaps using our knowledge of modern logic.[294] Thus, no attempt is made to modify the results found in Leibniz's notes.

This alternate way of looking at Leibniz's work results in the same conclusion: His construction is an ingenious anticipation of modern logic. But it also leads to the conclusion that the groundbreaking work on modern logic done by Boole and Frege in the 19th century—covered in the next two sections—was done independently of Leibniz's results.

Leibniz also envisioned a *calculus ratiocinator*[295] (computing evaluator) that, depending on interpretation and viewpoint, anticipates either a formal inference engine, a computer program, or a computer itself. In the third case, a calculator designed by Leibniz for the four arithmetic operations[296] can be viewed as one of the early steps in the development of computers.

The concept of calculus ratiocinator is closely connected with Leibniz's idea of a precise universal language he termed *characteristica universalis*[297] (universal characteristics). It was to support a precise encoding of the facts of the world. The calculus ratiocinator then would carry out an evaluation of these facts and verification of conclusions.

Thus, Leibniz anticipated the key idea of *artificial intelligence*[298] formulated in the 20th century. Indeed, he must be considered the founder of that discipline.

The above discussion has glossed over the variance among modern interpretations of the calculus ratiocinator and the characteristica universalis.[299] Nevertheless, there is universal agreement that Leibniz was a visionary for logic and its uses in mathematics and the everyday world.

We jump forward 150 years to Boole and Frege, who were not aware of Leibniz's results when they started their efforts. We first discuss Boole. His work marks the beginning of the 100 years of proof developments mentioned in the introduction of this chapter.

Boole: Reliable Logic Computation

Boole's work was motivated by the work of Augustus De Morgan (1806–1871), who defined basic concepts of logic such as propositions and rules of deduction.[300]

But Boole's approach was different from that of De Morgan, or for that matter, of Leibniz. While the latter

Augustus De Morgan.[301]

mathematicians strove for elaborate ways to formulate facts and relationships in logic statements, Boole aimed for a simple enough framework that allowed computations.

Indeed, Boole's most impressive achievement is the first complete calculus where mathematical operations reliably solve logic problems.[302]

AN INVESTIGATION

OF

THE LAWS OF THOUGHT,

ON WHICH ARE FOUNDED

THE MATHEMATICAL THEORIES OF LOGIC
AND PROBABILITIES.

BY

GEORGE BOOLE, LL.D.

PROFESSOR OF MATHEMATICS IN QUEEN'S COLLEGE, CORK.

LONDON:
WALTON AND MABERLY,
UPPER GOWER-STREET, AND IVY-LANE, PATERNOSTER-ROW.
CAMBRIDGE· MACMILLAN AND CO.
1854.

Left: George Boole.[303]
Right: Boole's The Laws of Thought, 1854[304]

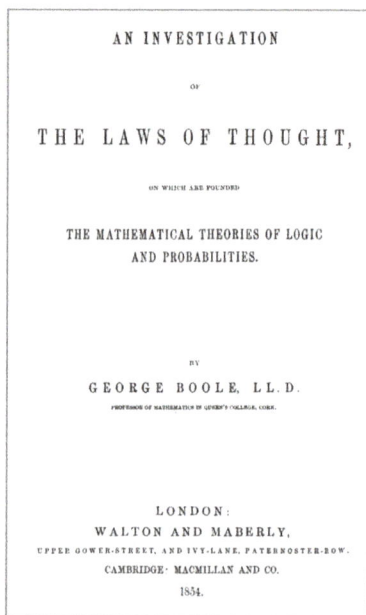

We couldn't state the key idea underlying the computational process more clearly and concisely than done by Boole in his book *The Laws of Thought:*[305]

"But as the formal processes of reasoning depend only upon the laws of the symbols, and not upon the nature of their interpretation, we are permitted to treat the above symbols x, y, z [which represent classes of objects that do or do not have certain properties] as if they were quantitative symbols of the kind above described.

"We may in fact lay aside the logical interpretation of the symbols in the given equations, convert them into quantitative symbols, susceptible only

of the values 0 *and* 1; *perform upon them as such all the requisite processes of solution; and finally restore to them their logical interpretation.*" [emphasis in the original]

Subsequently, Boole's system was simplified.[306] The resulting system is now called *Boolean algebra.*[307] Its formulas have an equivalent expression in the sentences of *propositional logic.*[308]

Frege: Definitions for Logic

In the groundbreaking book *Begriffsschrift*[309] (Treatise of Concepts), Frege created definitions of clarity and precision for logic that are still in use today.[310] Translated into English, the subtitle summarizes the key idea of the book as follows: *A language of formulas of pure thought, inspired by arithmetic.*

Left: Gottlob Frege.[311]
Right: Frege's Begriffsschrift, 1879.[312]

For the first time, Frege's definitions allowed rigorous modeling of the logic structure of complex situations, whether occurring in mathematics or the world. Fourteen years later, in 1893, Frege dared to construct in *Grundgesetze der Arithmetik*[313] (Fundamental Laws of Arithmetic) a foundation of arithmetic relying solely on logic. In-between, he had discussed a number of philosophical questions about arithmetic in a related book[314] where, in his words,[315] "[he] had tried to show that the arithmetic is a branch of logic which need not base the motivation for proofs on any experience or image."

In Grundgesetze der Arithmetik, this claim was to be proved by showing that the simplest laws of arithmetic could be established with just the tools of mathematical logic of the Begriffsschrift. The book represents a monumental effort where hundreds of logic diagrams prove the claims.

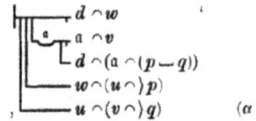

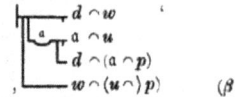

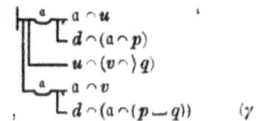

Left: Frege's Grundgesetze der Arithmetik, 1893[316]
Right: Logic diagrams of Grundgesetze der Arithmetik, detail of p. 76 left column.

As Frege was completing the book, he received a letter from Bertrand Russell (1872–1970) that pointed out a devastating inconsistency in a basic assumption of the book. Frege's response was a painstaking analysis in a 13-page appendix to discover the root of the difficulty and suggest possible remedies. Unfortunately, at the end he had to admit that he did not have a satisfactory solution.[317]

Bertrand Russell.[318]

The problem concerned Frege's definition of a *set*, which generally is a collection of *elements*. For example, we may declare *primes* to be the set whose elements are the prime whole numbers.

The concept of set is essential when one constructs any area of mathematics using logic: Mathematical claims can then be rephrased as logic statements claiming certain elements to be in a set, or a certain set to be contained in another set, and so on.

Frege defined a set to be any collection that can be defined by a logic condition. For example, the set of prime numbers n can be defined by imposing the logic condition that n must be a natural number that cannot be factored in a nontrivial way. Unfortunately, Frege's definition admits sets with contradictory properties,[319] as discovered by Russell.

We move forward to the beginning of the 20th century. At that time, a historic battle within mathematics started about the validity of mathematical assumptions and proof techniques.

We saw an early skirmish of that battle in Chapter 4, where Cantor defended his revolutionary ideas about infinity against Kronecker's demand of finitism. But now a full-blown battle unfolded over choice in assumptions and proofs.

The story begins with David Hilbert (1862–1943), who arguably was the most prominent mathematician of the late 19th and early 20th century. His interests were universal, and his results had far-reaching impact.[320] Here, we will focus on a set of problems he posed right at the beginning of the 20th century.

Hilbert: 23 Problems

In 1900, Hilbert posed 23 unsolved problems that were of eminent importance. Indeed, they motivated a large part of the mathematical research of the 20th century.[321]

Among them was Cantor's continuum hypothesis, which says that there is no set whose cardinality lies strictly between that of the natural numbers and the real numbers.

Put differently, it claims that the real numbers constitute a smallest uncountable set.

David Hilbert.[322]

As part of that hypothesis, Hilbert asked that a certain result called the *well-ordering theorem* be proved, where *well ordering* is defined as follows: A set with a given ordering of the elements is *well ordered* if every nonempty subset has a smallest element.[323]

The real numbers with their natural ordering are not well ordered. For example, take the subset of real numbers greater than 0. It does not have a smallest number, since for any $x > 0$ of the set, $\frac{x}{2}$ is also in the set. Thus, there cannot be a smallest $x > 0$.

The *well-ordering theorem* says that, for every set, there exists some ordering such that the set becomes well ordered. In particular, the theorem implies that there is an ordering of the set of real numbers such that it becomes well ordered, a strange claim at the time since nobody had a clue how to do this.

Zermelo: Axiom of Choice

In 1904, Ernst Zermelo (1871–1953) set out to prove the well-ordering theorem. For this task, he invented an axiom that later was

called the *axiom of choice*. At the time, it looked to be a totally self-evident axiom, and Zermelo had no compunction in defining and using it. We describe the axiom next.

Imagine that a set is represented by a basket with a lid. Any elements of the set are represented by items in the basket. Thus, an empty basket corresponds to the empty set.

Consider a long row of baskets of the above kind. Somebody assures us that each basket contains at least one item. We are asked to get exactly one item from each basket and set it down next to the basket, as sort of demonstration that the basket indeed isn't empty.

This is easily done. We step up to each basket, open the lid, grab an item, and put it down next to the basket. The task may be time-consuming if

Ernst Zermelo.[324]

Basket with lid.[325]

the row of baskets is long, but in principle we can do this.

Now make the row of baskets longer and longer. At the same time, imagine the baskets to be labeled, say with the natural numbers 1, 2, 3, ... , with no limit.

We are still assured that each basket is nonempty. We are asked to carry out the above process, but with the additional condition that we are to assign the label of each basket to the removed item.

Thus we get $item_1$, $item_2$, $item_3$, ... Of course, we cannot complete the process in finite time. But we can imagine doing it forever, and in the process will achieve the goal.

We come to an even bigger collection of baskets. Each basket has again a label, but there are so many baskets that we cannot arrange them in a row. Instead, there is an arbitrarily large collection L of labels, and for each label x in L, denoted by $x \in L$, there is exactly one basket labeled with x.

Once more, we are told that each basket is nonempty, and we are asked to remove one item from each basket and set it down next to the basket. We can readily do this for a given basket; say, we remove $item_x$ from basket x.

But we are hard-pressed to describe a removal process that for any collection L of indices will complete the task even in infinite time.

The *axiom of choice* states that the desired removal is not only possible, but can be carried out in one step regardless of the size of the collection L.

So given nonempty baskets labeled by L, in one fell swoop we obtain $item_x$ from basket x, for each $x \in L$. For the case of countable L—for example, in the above case of labels 1, 2, 3, . . .—the axiom becomes the less demanding *axiom of countable choice.*[326]

The axiom of choice had been implicitly employed prior to Zermelo. For example, Cantor used it in the construction of certain infinite sets without recognizing that he was invoking something outside standard set theory.

Zermelo was first to realize, in 1904, that the axiom needed explicit formulation. To him, the axiom of choice seemed of self-evident reasonableness and simplicity. Accordingly, he felt free to use it in his proof of the well-ordering theorem.

In 1908, Zermelo proposed axioms of set theory that were later modified by Abraham Fraenkel (1891–1965). The resulting *Zermelo-Fraenkel set theory* included the axiom of choice. That system is denoted by *ZFC*, where the

Adolf Abraham Halevi Fraenkel.[327]

"C" signifies use of the axiom of choice. It was soon determined that the axiom of choice had profound consequences. Accordingly, Zermelo-Fraenkel set theory without the axiom of choice became of interest; it is denoted by *ZF*.

Banach and Tarski: Paradox

In 1924, Stefan Banach (1892–1945) and Alfred Tarski (1901–1983) proved the following astonishing result, now called the *Banach-Tarski paradox:*[328]

The axiom of choice of *ZFC* may be used to divide a solid 3-dimensional sphere into a great many small pieces,

Banach-Tarski Paradox: A solid ball is turned into two solid balls of the same size.[329]

and then reassemble these very pieces and obtain two solid 3-dimensional spheres, each with the *same* diameter as the original one. In short, the axiom allows a doubling of material.

Left: Stefan Banach.[330]
Right: Alfred Tarski.[331]

The paradox proved that the axiom of choice has strange and counterintuitive effects, and thus destroyed the notion that the axiom could be justified by appeal to physical everyday processes such as removing items from baskets.

Six years earlier, in 1918, L. E. J. Brouwer (1881–1966) had raised a more fundamental objection to proof processes based just on *ZF*, and thus had started a new philosophy of mathematics called *intuitionism*.[332]

At that time, Brouwer was known for several profound results in *topology*, the branch of mathematics concerned with properties of space that are independent of stretching and bending.[334]

L. E. J. Brouwer.[333]

An example is his *fixed-point theorem*,[335] which stands out in a long list of fixed-point results due to its widespread use in various fields of mathematics.

An illustrative example application[336] of the theorem is as follows. Take two sheets of paper of same size. Let one page cover the other one. Then each point of the bottom page corresponds to a unique point of the top page.

Now crumple up the top page in any way and position it so that it does not extend beyond the boundary of the bottom page. Then there is a point of the crumpled page such that the corresponding point of the flat page lies vertically below.

Brouwer: Intuitionism

Brouwer's intuitionism[337] views mathematics as a constructive process. In particular, a mathematical object can be claimed to exist only if it can be constructed.

As an example, suppose we have a set S that we guess to be not empty. We wish to remove one element from S for additional mathematical manipulation. Unfortunately, we somehow fail to assemble direct logic arguments that provide the desired element.

At this point, the axiomatic method allows an indirect path to our goal. We assume that S is empty and, with appropriate arguments of logic, aim for a contradictory result that invalidates that assumption.

Suppose we succeed in this endeavor. Thus, we have shown that the statement "S is empty" is false. Using Aristotle's *law of the excluded middle*,[338] there is only one alternative, which means that the statement "S is not empty" is true.

Thus, S contains at least one element. We arbitrarily pick one such element and use it in further mathematical manipulation, as desired.

The intuitionist goes along with this method right up to the point where the assumption "S is empty" results in a contradiction.

But the next step—claiming that S is not empty and extracting one item from S—is not permissible. Indeed, the intuitionist obtains such an item only by a suitable construction and not simply by the fact that the claim "S is empty" has produced a contradiction.[339]

In 1921, Hermann Weyl (1885–1955) clarified and expanded Brouwer's ideas.[340]

Weyl eventually became one of the most influential mathematicians of the 20th century. He achieved outstanding results in several branches of mathematics including number theory, and in theoretical physics.

Hermann Weyl.[341]

During the second half of the 1920s, Weyl began to realize that the restrictions imposed by intuitionism would cripple mathematics. At that same time, Hilbert and Brouwer engaged in a monumental struggle about the future direction of mathematics. The war culminated in an ugly battle about

membership on the editorial board of the leading mathematical journal *Mathematische Annalen.*[342]

On the one hand, Hilbert wanted to ensure that the editorial board accepted the axiomatic method, in which rules of logic derive results from basic axioms.

These rules allow simple steps such as claiming use of an element of a set that has been proved to be nonempty by contradiction—see the earlier example—as well as much more complicated creation and manipulation of functions or sets based on the axiom of choice.

In fact, by the end of the 1920s, the axiom of choice was fully accepted as part of Zermelo-Fraenkel set theory, despite disturbing results such as the Banach-Tarski paradox. As stated earlier, that version is denoted by *ZFC*, and the version without the axiom of choice by *ZF*. Today, the system *ZFC* is considered the foundation for most of mathematics.[343]

On the other hand, Brouwer wanted to see all such steps ruled out unless supported by constructive arguments. Acceptance of that severe condition would have invalidated many results of Hilbert and others, and would have precipitated a historic destruction of mathematics.

No wonder Hilbert and like-minded mathematicians waged a ferocious battle against Brouwer's demand that mathematics be based on the concepts of intuitionism.

Fortunately for mathematics, Hilbert's side won, and today nobody seriously questions whether non-constructive results are acceptable.

Looking back now, almost 100 years later, Brouwer's life is seen to be a tragedy: Here was a brilliant mathematician who created groundbreaking results in topology during the four years spanning 1909–1913 and thus had started an illustrious mathematical career.

Hilbert was so impressed that in 1912 he helped Brouwer obtain a regular academic appointment at the University of Amsterdam.[344]

With financial support secured, Brouwer embarked on development of the philosophy of intuitionism to the exclusion of everything else. This effort spanned decades of his life, but eventually turned out to be futile.

We close the discussion of intuitionism with a quote by Weyl, who by 1949 had become disillusioned with intuitionism. His statement rejects intuitionism, but also acknowledges Brouwer's lofty goals:[345]

"Mathematics with Brouwer gains its highest intuitive clarity. He succeeds in developing the beginnings of analysis in a natural manner, all the time preserving the contact with intuition much more closely than had been done before.

"It cannot be denied, however, that in advancing to higher and more general theories the inapplicability of the simple laws of classical logic eventually results in an almost unbearable awkwardness. And the mathematician watches with pain the greater part of his towering edifice which he believed to be built of concrete blocks dissolve into mist before his eyes."

We go back to the beginning of the 1910s, when Alfred North Whitehead (1861–1947) and Russell proposed a novel construction of mathematics based on logic.

In 1893, Frege had attempted such a construction in the Grundgesetze der Arithmetik, but had failed due to a contradictory definition of sets.

Whitehead and Russell: Principia Mathematica

During the period 1910-1913, Whitehead and Russell published the *Principia Mathematica*, a landmark achievement that in three volumes defined a foundation of mathematics.[346]

Their approach was based on the long-held belief that all of mathematics could be constructed from elementary principles: One just had to find the right way to define the fundamental concepts, determine a reliable construction method, and then all of mathematics could be built.

For Whitehead and Russell it seemed obvious, just as it did for Frege, that these fundamental principles were to be found in the logic created in the 19th century.

PRINCIPIA MATHEMATICA

BY

ALFRED NORTH WHITEHEAD, Sc.D., F.R.S.
Fellow and late Lecturer of Trinity College, Cambridge

AND

BERTRAND RUSSELL, M.A., F.R.S.
Lecturer and late Fellow of Trinity College, Cambridge

VOLUME III

Cambridge
at the University Press
1913

Above: Alfred North Whitehead.[347]
Right: Principia Mathematica, Vol. III, 1913.[348]

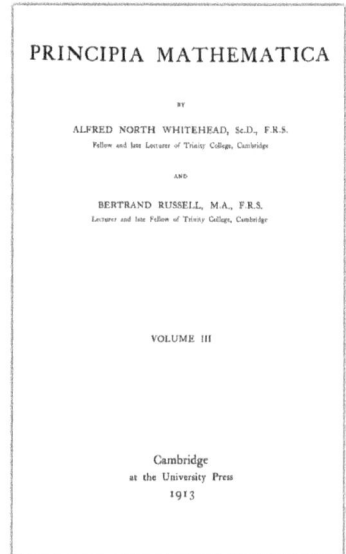

Whitehead and Russell worked carefully to avoid the conundrum of Frege, where the definition of sets had introduced a fatal contradiction.

The painstaking approach started with the basic rules of propositional logic and then built up result after result through an intricate network of logic formulas.[349] Part of the construction was a complicated hierarchy of sets.

After 379 pages of volume I, a logic statement equivalent to the arithmetic result $1 + 1 = 2$ was proved. Due to that slow pace, the three volumes covered only set theory, cardinal numbers, ordinal numbers, and real numbers.

But it seemed evident that the construction method, suitably continued, would produce a large portion if not all of mathematics.

A second edition was published in 1925, and an abbreviated version in 1997, showing that interest in the work continued over decades.

Comments ranged from suggestions for modifications to fundamental criticism. We will skip that extensive material[350] and instead focus on two aspects: the *completeness* and *consistency* of systems of axioms. The definitions of these two terms are as follows:

A system of axioms is *complete* if every true statement can be proved in the system. It is *consistent* if it does not contain a contradiction.[351]

The *Principia Mathematica* starts with propositional logic, which is complete and consistent. At the time, it was assumed that the two properties would be maintained by suitable construction of results.

That aspect, indeed the general problem of completeness and consistency of mathematics, was taken up by Hilbert. He wanted to place all of mathematics on a reliable foundation where completeness and consistency were proved throughout.[352]

Hilbert: Completeness and Consistency

In 1880, the physician and physiologist Emil du Bois-Reymond (1818–1896) argued in a speech before the Berlin Academy of Sciences that the Latin maxim *ignoramus et ignorabimus* (we do not know and we will not know) applies to certain parts of the world.[354]

In particular, mankind would never understand the nature of matter and force, the origin of motion, and the

Emil du Bois-Reymond.[353]

origin of simple sensations. Bold predictions are often wrong. This certainly applies here: Matter, force, and motion have been explained by Einstein's theory of relativity and quantum physics; and

simple sensations have been investigated by brain science, along with many other interactions of mind and body with the external world.

In 1930, Hilbert delivered a celebrated address to the Society of German Scientists and Physicians where he argued passionately against the mentality of *ignorabimus:*[355]

"We must not believe those, who today, with philosophical bearing and deliberative tone, prophesy the fall of culture and accept the *ignorabimus*. For us there is no *ignorabimus*, and in my opinion none whatever in natural science.

"In opposition to the foolish *ignora-bimus* our slogan shall be: Wir müssen wissen – wir werden wissen! (We must know – we will know!)"

The famous statement "Wir müssen wissen – wir werden wissen" eventually was engraved on Hilbert's tombstone.

Part of Hilbert's tombstone: Wir müssen wissen – wir werden wissen (We must know, we will know).[356]

Hilbert's arguments against Du Bois-Reymond's predictions were well justified. Hadn't several problems open since antiquity been solved in the 19th century? Didn't Frege design the Begriffsschrift, which allowed accurate encoding of all facts in logic? Weren't the axioms of Zermelo-Fraenkel a precise formulation of set theory? Didn't Whitehead and Russell show that basic concepts of numbers and related operations could be developed just starting with logic?

So Hilbert decided that there was the opportunity—indeed the duty—to answer once and for all the fundamental questions of mathematics. In particular, the nagging uncertainty about completeness and consistency of systems, some used for thousands of years, had to be removed.

In the 1920s, and thus before the talk cited above, Hilbert had already started a project, now called *Hilbert's Program*, that was to create a secure foundation for all of mathematics.

His program had a number of goals.[357] The two most important ones were proof of *completeness* and *consistency* of all axiomatic systems.

The process proving completeness and consistency of mathematical systems obviously would use mathematical arguments. How could Hilbert be sure that these arguments themselves were not flawed?

To preclude that disastrous possibility, he outlined a very restricted mathematics called *finitary* that unquestionably could be used to prove completeness and consistency for general mathematical systems.[358]

With that finitary proof machinery established, he and others set out to process the existing mathematical systems one by one, each time searching for proofs of the desired features. Early successes for simple mathematical systems supported Hilbert's optimism that the program could indeed be carried out. For example, the theory of only addition of natural numbers and of multiplication of the positive integers was proved to be complete and consistent.[359]

But then progress came to a halt: In 1931, Kurt Gödel (1906–1978), at age 25, published devastating results that limited what could *ever* be ascertained by mathematical arguments. That work and other results in logic prove Gödel to be one of the greatest logicians of all time.

Kurt Gödel, ca. 1926.[360]

Gödel: Incompleteness

Gödel investigated what could ever be established for a given axiomatic system when all arguments are based just on those axioms and do not use any external assumptions.

That effort produced the following two results in 1931, now called *incompleteness theorems*, that forever limit what can be proved that way:[361]

First incompleteness theorem: For any formal mathematical system that contains a certain amount of elementary arithmetic, there are statements composed in the language of the system that can be neither proved nor disproved in that system.

Example systems are Zermelo-Fraenkel set theory without axiom of choice (*ZF*) or with that axiom (*ZFC*). For either system, there are statements formulated in the language of that system that cannot be proved or disproved. If such statements are added as axioms, then there are new statements that cannot be proved or disproved. Thus, there is no remedy of incompleteness of the system.

Second incompleteness theorem: For any consistent system that contains a certain amount of elementary arithmetic,[362] consistency of the system cannot be proved in that system.

This second result also applies to *ZF* and *ZFC*. That is, one cannot possibly prove consistency in those systems.

Taken together, the two results dealt a fatal blow to Hilbert's program:[363] There was no more hope that any mathematical system of reasonable sophistication could be proved complete or consistent, just using the axioms of the system itself.

In particular, for *ZFC* there could never be a proof of consistency constructed within that system. Yet, that theory was, and still is, the foundation of most of mathematics.

So after centuries of construction, mathematicians have created a foundation for almost all of mathematics that might be faulty. This is a mind-boggling conclusion.

Now *ZFC* has been used almost 100 years, and no case of inconsistency has surfaced. If we were to argue like Wallis of the 17th century,[364] we would say that there have been plenty of experiments to support the belief that there is no inconsistency.

That sort of argument for validity was rejected when Wallis used it to justify his results. Yet exactly this argument is made today regarding the most fundamental part of mathematics.

There are two alternatives that avoid the consistency problem at the foundation of mathematics.

First, we could proceed like physicists, who still use Newton's model of the world instead of Einstein's theory of relativity or quantum physics initiated by Max Planck (1858–1947), as long as the errors are so small as to be irrelevant.

Analogously, we could have a construction of mathematics guided by the philosophy of Wallis where all results are verified by experiments and used until a serious error surfaces. But what a confusing world it would

Max Planck.[365]

be! So no mathematician would agree to such a course of action.

Second, we could confine mathematics to finiteness, as proposed by Kronecker, and then handle complicated concepts like $\sqrt{2}$ and π by approximations.

Precisely this restriction to finiteness is happening in modern computation, almost without exception.

But a general restriction to finiteness would remove soaring ideas such as Cantor's construction of the infinite cardinal and ordinal numbers. Indeed, it would reduce mathematics to a drab world.

So, mathematicians have accepted a compromise. They tolerate the potential for inconsistency in the foundation of mathematics, but from then on carry out a construction with unassailable proofs and thus can build result upon result.

The conclusions are *relatively valid*: If the foundation is consistent, then all results hold. If an inconsistency ever surfaces in the foundation, mathematicians hope—indeed anticipate—that the problem can somehow be remedied by a suitable modification that does not destroy the entire edifice. They also accept that the method cannot establish all true results as indeed true, an unavoidable shortcoming established by Gödel's first incompleteness theorem.

For the next step in the history of mathematical proof, we need the concept of *independence of axioms*, defined as follows.

Let S be a consistent set of axioms, and define A to be some other axiom. Then axiom A is declared to be *independent* from S if the following two systems are consistent: S with A added, and S with the negation of A added.[366]

An important question since the definition of ZF was: Assuming that the axioms of ZF are consistent, are the axiom of choice and the continuum hypothesis independent of ZF?

Gödel partly proved this. In 1938, he showed the following: If ZF is consistent, then ZF with the axiom of choice added is consistent as well. In 1940, he proved the analogous result for the continuum hypothesis.[367]

But then no further progress was made. Indeed, many others tried to answer the open question about consistency when the negation of the axiom of choice or of the continuum hypothesis is added.

The problem seemed so difficult that young mathematicians were counseled not to work on that seemingly hopeless problem. Fortunately, one mathematician ignored that advice.

Cohen: Independence of Choice

In 1963, at age 29, Paul J. Cohen (1934–2007) published the following result:[368]

If ZF is consistent, then addition of the negation of the axiom of choice or of the negation of the continuum hypothesis results in another consistent system.

That proof plus Gödel's earlier conclusion established the long-sought
result that the axiom of choice and
the continuum hypothesis are independent of *ZF*.

Gödel sent a comment to Cohen, a
draft of which has survived. It says,

"Let me repeat that it is really a
delight to read your proof of the
ind[ependence] of the cont[inuum]
hyp[othesis]. I think that in all essen-

Paul J. Cohen.[369]

tial respects you have given the best possible proof & this does not
happen frequently. Reading your proof had a similarly pleasant
effect on me as seeing a really good play."[370]

Cohen's proof was based on *forcing*,[371] a method he invented for
proving consistency and independence of results. The method has
become a standard tool for work on the foundation of mathematics.

Using forcing, he also showed that the independence result for
the continuum hypothesis remains correct when the larger system
ZFC, which is *ZF* plus the axiom of choice, is used.

That is, if *ZFC* is consistent—which is the case if *ZF* is consistent—
then the continuum hypothesis is independent of *ZFC*.

We thus can declare either *ZFC* with the continuum hypothesis
added or with the negation of that hypothesis added to be our
foundation of mathematics, and are assured that either choice is
consistent if *ZF* is consistent. Which of the two possibilities should
be selected for the mathematics of the future? There is substantial
debate about this, with no easy resolution in sight.[372]

That debate once more shows that development of the foundation
of mathematics is an ever ongoing process.

In all informal discussions about mathematics, we have used the
image of a building where the load-bearing columns in the

basement correspond to the foundation of mathematics. A different and richer image was proposed in 1948 by Nicolas Bourbaki;[373] the name is a collective pseudonym of a group of mathematicians who have attempted a comprehensive coverage of mathematics:

"Mathematics is similar to a large city where suburbs grow into the surrounding land and where the center is periodically rebuilt, each time according to a clearly defined plan and according to a new, more impressive order."[374]

But even this expanded picture does not capture the richness of mathematical developments. In the terminology of the Bourbaki paper, we encounter next the construction of an extraordinary bridge connecting two distant suburbs of the city of mathematics.

Wiles: Proof of Fermat's Last Theorem

In 1637, Fermat claimed the following, now known as *Fermat's last theorem*:[375]

The equation $x^n + y^n = z^n$, with n any natural number greater than 2, has no solution where x, y, and z are natural numbers.

For the case $n = 2$, there are of course solutions with natural numbers; a collection of such solutions, now called Pythagorean triples, was already created in ancient Babylon, as cited earlier in this chapter.

Fermat wrote in the margins of one of his books that he had a proof.[376] But it is most likely that he was mistaken in this assessment, given the complexity of the eventual proof.

Over centuries, Fermat's claim was proved[377] for ever larger values of n. Computers of the 20th century allowed a massive numerical evaluation that confirmed Fermat's claim to be correct up to $n = 4,000,000$.

In 1995, more than 350 years after Fermat posted the claim, Andrew John Wiles (1953–) published a proof of the theorem.[378]

It was a stunning achievement. The seeds for the proof were planted in 1955, when Yutaka Taniyama (1927–1958) and Goro Shimura (1930–) created an utterly daring conjecture: Two mathematical areas that to all appearances were completely different, were claimed to be fundamentally linked.[379]

The proposed link seemed preposterous. In terms of Bourbaki's city with suburbs, the conjecture envisioned a bridge connecting two far-apart suburbs. That conjecture, unlikely as it seemed, turned out to be correct

Andrew John Wiles.[380]

and is now called the *modularity theorem*.[381] Wiles proved a special version of the conjecture that was sufficient to establish Fermat's last theorem.

The story is too complex to be covered here, but has been well described.[382] Building on Wiles's work, other mathematicians proved the remaining portion of the modularity theorem during 1996–2001.

Summary

The chapter has traced the history of the concept of mathematical proof, beginning with Babylonian results based on geometry and groundbreaking efforts of the ancient Greeks.

The next milestone was Leibniz, who anticipated the concepts of modern logic so essential for reliable proofs. Boole and Frege then created key parts of that machinery.

Late in the 19th century and continuing into the 20th century, mathematicians tried to create a precise foundation of mathematics and reliable methods for proofs.

That process involved Frege, Zermelo, Fraenkel, Whitehead, and Russell. It essentially ended up with Zermelo-Fraenkel set theory as the basis for most of mathematics.

A struggle about the use of axioms and the interpretation of mathematical steps pitted Brouwer and Weyl, the proponents of intuitionism, against mathematicians led by Hilbert.

The latter group freely used axioms to build astonishing results. Hilbert also launched a drive to put all of mathematics on a permanent, unassailable foundation.

That effort collapsed when Gödel proved that this goal will forever be unattainable. In particular, consistency of the foundation axioms will always be in doubt. Gödel also showed that two advanced axioms—the axiom of choice and the continuum hypothesis—can be added to ZF without fear of inconsistency, assuming that ZF itself is consistent.

Cohen proved the very difficult additional result that the negated versions of these axioms can also be added without loss of consistency.

We thus have wide-open choices about the foundation of mathematics, where the axiom of choice or the continuum hypothesis can be added to ZF or declared to be not valid. All such choices are allowed under the assumption that ZF is consistent.

At this time, the addition of the axiom of choice is fully accepted, and the resulting system ZFC is generally used. But addition of the continuum hypothesis or its negation is still being debated.

Lastly, Wiles's proof of Fermat's last theorem shows that seemingly disparate areas of mathematics are sometimes closely connected. This result has motivated the current search for profound links joining widely different regions of mathematics.[383]

_____

You have reached the end of the book. The epilogue has some additional thoughts.

Epilogue

If you've ever been frustrated by math, we hope the material has helped you: first, to see that famous mathematicians struggled over decades or even centuries to achieve clarity on fundamental issues; second, to gain insight into mathematical concepts from that evolutionary process; third, to overcome negative, indeed damaging, feelings about mathematics since they are just a consequence of inappropriate teaching and have nothing to do with your abilities.

Further advice: If you haven't read the extensive notes, you may pick chapters that have helped you gain insight, read the related notes, and follow up by investigating the cited material. It will deepen your understanding.

I wish I had had a book like this as an engineering student, when I often experienced misgivings about mathematical claims. Much later, as a professor, I realized my doubts were well-justified. The thought surfaced: Could I help students facing similar difficulties?

I am convinced that I cannot influence how teachers teach, as I learned during occasional futile discussions. But then, why not turn directly to students and help them understand the fundamental concepts of mathematics? That thought is what motivated this book.

Notes

Throughout, "Wikipedia" refers to the English version unless another language is explicitly listed.

Chapter 1 Introduction

1. For further exploration of the history of mathematics, we have found the 2,535-page 4-volume set *The World of Mathematics* by James Roy Newman (1907–1966) ([Newman, 1956]) to be very helpful, as well as a number of books by Florian Cajori (1859–1930). The website archive.org supplies the 4-volume set [Newman, 1956] as well as a number of Cajori's books, for example [Cajori, 1918], [Cajori, 1919a], [Cajori, 1919b], and [Cajori, 1928].

Chapter 2 Numbers

2. See Wikipedia "Ocean."

3. Source: `https://en.wikipedia.org/wiki/Georg_Cantor#/media/File:Georg_Cantor2.jpg`. "Georg Cantor2" by Unknown - `http://i12bent.tumblr.com/post/3622180726/georg-cantor-german-mathematician-and-philosopher`. Licensed under Public Domain via Commons.

4. [Rudman, 2007] covers the first 50,000 years of mathematics in clear and concise detail. The first part of this section relies on that material.

5. p. 54 [Rudman, 2007].

6. p. 68 [Rudman, 2007].

7. The number 1 is not considered a prime. If it was, then the equa-

tion $1 = 1 \cdot 1 = 1 \cdot 1 \cdot 1$ would establish that the representation of any natural number as a product of primes was not unique, an undesirable conclusion.

8. Proof: If there are only a finite number of primes, then multiply them all together and add 1. The resulting number cannot be divided by any of the primes in the collection, a contradiction of the claim that the given primes can produce all numbers.

9. Proof: First establish that, if a prime p divides $a \cdot b$, then p divides a or b.
Assume the opposite. Thus, for some integer m, $a \cdot b = p \cdot m$, and p does not divide a or b. Pick an instance with p as small as possible, and subject to that, a as small as possible. If $a > p$, the equation $(a - p) \cdot b = p(m - b)$ constitutes a smaller instance, a contradiction. If $a = p$ or $a = 1$, then p divides a or b, another contradiction. Thus, $1 < a < p$. Now a is the product of some prime, say q, and some $c \geq 1$. If q divides m, then $(a/q) \cdot b = p \cdot (m/q)$ is a smaller instance. This leaves the case where q does not divide m. Since q cannot divide the prime p, the equation $q \cdot (c \cdot b) = p \cdot m$ establishes a smaller instance, with q playing the role of p. Thus, all cases result in a contradiction, and the claim must hold.
We are ready to prove the unique factorization claim by induction. The result is immediate if the factored number is a prime. Otherwise, take any two factorizations. Select any prime factor of the first factorization. Split the second factorization arbitrarily into two parts. Then by the above result and induction, one of the two parts must contain the prime. Remove the prime from both factorizations, and repeat the argument. Thus, both factorizations are shown to be identical.

10. See Wikipedia "Fundamental Theorem of Arithmetic." The entry identifies the results by Euclid that directly imply the theorem.

11. [Sautoy, 2003].

12. Source: https://en.wikipedia.org/wiki/Euclid#/media/File:Euklid.jpg. "Euclid of Megara" (lat: Evklidi Megaren), Panel from the Series "Famous Men," Justus of Ghent, ca. 1474, Panel, 102cm x 80cm, Urbino, Galleria Nazionale delle Marche. This picture is meant to represent the famous mathematician Euclid of Alexandria, who in medieval times was wrongly identified with Euclid of Megara, the disciple of Socrates.

13. [Fritz, 1945].

14. $\sqrt{2}$ is the number that, when multiplied with itself, results in 2. In general, the kth root of a number n, denoted by $\sqrt[k]{n}$, is the number that, when multiplied together k times results in n.

15. Proof that $\sqrt{2}$ is irrational, using Euclid's result that each integer is a product of unique primes: Suppose $\sqrt{2}$ is rational, so for some integers a and b, $\sqrt{2} = \frac{a}{b}$, or equivalently $b \cdot \sqrt{2} = a$. Squaring each side, we get $b^2 \cdot 2 = a^2$. Now a and b are the unique product of some primes. Thus, by the uniqueness of the representations of a, a^2, b, and b^2, each prime factor of a^2 as well as of b^2 occurs an even number of times. Then the factorization of $b^2 \cdot 2$ has the prime 2 occurring an odd number of times, while it must occur in a^2 an even number of times, a contradiction. The proof is readily adapted to prove the following for any integer $k \geq 2$ and any positive integer n: If $\sqrt[k]{n}$ is not an integer, then it is not rational.

16. Source: https://www.math.ubc.ca/~cass/Euclid/ybc/ybc.html. Permission for use kindly granted by William A. Casselman, photographer and copyright holder. The clay tablet is part of the Yale Babylonian Collection. Provenance unknown, dated ca. 1800–1600 BCE. Purchased around 1912 by an agent of J. P. Morgan, who contributed it to Yale University as part of the foundation of its Babylonian Collection. For details about the numbers of the tablet and the computation of $\sqrt{2}$, see Wikipedia "Square root of 2."

17. Dedekind motivates his search for a construction of the irrational numbers as follows ([Dedekind, 1872]):
"Just as the negative integers and fractions of integers are produced by free creation, and just as the laws of computations must and can be based on the computations for natural numbers, in like manner one has to strive that the irrational numbers as well are defined entirely via the rational numbers. Only the How? remains the question."

18. Source: https://en.wikipedia.org/wiki/Richard_Dedekind#/media/File:Richard_Dedekind_1900s.jpg. "Richard Dedekind 1900s" by Unknown (Mondadori Publishers). Licensed under Public Domain via Commons.

19. The precise construction is as follows: Consider the rational numbers placed into a sorted sequence. To construct a single irrational number, cut the sorted sequence at some point, getting a sequence A below the cut and a sequence B above the cut. Select

the cut in such a way that A has no greatest element. Then there is a unique real number that lies between A and B. If B has no least element, then the constructed number is irrational; otherwise it is rational. The figure below shows the construction of $\sqrt{2}$.
Consider the process repeated for all possible cuts. Then all real numbers are constructed.

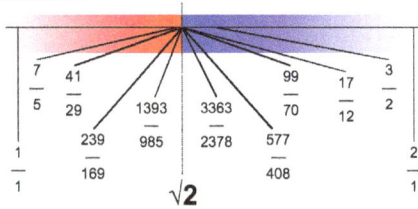

$\sqrt{2}$

(Source: https://commons.wikimedia.org/wiki/File:Dedekind_cut-_square_root_of_two.png#/media/File:Dedekind_cut-_square_root_of_two.png. "Dedekind cut - square root of two" by Hyacinth - Own work. Licensed under Public Domain via Commons.)

20. More precisely, the polynomial has the form $a_n \cdot x^n + a_{n-1} \cdot x^{n-1} + \cdots + a_1 \cdot x^1 + a_0$, where n is some positive integer and all a_i are integers with $a_n \neq 0$.

21. The integers are algebraic since for any integer a, the polynomial $x - a$ has a as unique root. The rational numbers are algebraic numbers as well since, for integers a and b, $x = \frac{a}{b}$ is the unique root of the polynomial $b \cdot x - a$.

22. Source: https://commons.wikimedia.org/wiki/File:Leonhard_Euler_2.jpg. "Leonhard Euler 2" by Jakob Emanuel Handmann - 2011-12-22 (upload, according to EXIF data). Licensed under Public Domain via Commons.

23. Source: https://en.wikipedia.org/wiki/Joseph_Liouville#/media/File:Joseph_liouville.jpeg. "Joseph liouville" by http://www.math.sunysb.edu/. Licensed under Public Domain via Commons.

24. [Erdös and Dudley, 1983].

25. See Wikipedia "Transcendental number."

26. The most famous equation constructed by Euler is $e^{i\pi} + 1 = 0$. The constant $e = 2.7182\ldots$ is now called the *Euler number*. The constant $\pi = 3.1415\ldots$ links diameter d and circumference c of a circle by $c = d \cdot \pi$. The physicist Richard Feynman called Euler's

equation "our jewel" and "the most remarkable formula in mathematics;" see Wikipedia "Euler's formula."

27. See Wikipedia "Complex Plane."

28. Addition of complex numbers translates to addition of vectors, and multiplication is represented by a certain rotation of vectors. See Wikipedia "Complex Number."

29. Source: `https://en.wikipedia.org/wiki/Complex_number` `#/media/File:Complex_number_illustration.svg`. "Complex number illustration" by Wolfkeeper. Licensed under CC BY-SA 3.0 via Commons.

30. See Wikipedia "e (mathematical constant)."

31. Source: `https://en.wikipedia.org/wiki/Jacob_Bernoulli#` `/media/File:Jakob_Bernoulli.jpg`. "Jakob Bernoulli" by Niklaus Bernoulli (1662-1716). Licensed under Public Domain via Commons.

32. Suppose an account holding \$1 earns 100% interest in a year. If the interest is credited at the end of the year, the account will grow to \$2. When the interest is credited twice a year at $\frac{100}{2} = 50\%$, the account will grow to \$2.25. Now suppose the account is credited n times a year, with rate $\frac{100}{n}\%$. As we consider ever larger values of n, the account value at the end of the year approaches, and in the limit reaches, $e = 2.7182\ldots$.

33. See Wikipedia "e and pi are transcendental."

34. Source: `https://en.wikipedia.org/wiki/Charles_Hermite#` `/media/File:Charles_Hermite_circa_1901_edit.jpg`. "Charles_Hermite_circa_1901" by Unknown, derivative work: Quibik (talk) - Charles_Hermite_circa_1901.jpg. Licensed under Public Domain via Commons.

35. Source: `https://en.wikipedia.org/wiki/Ferdinand_von_Lind` `emann#/media/File:Carl_Louis_Ferdinand_von_Lindemann.jpg`. "Carl Louis Ferdinand von Lindemann" by Unknown - `http://ww` `w.math.uha.fr/Pi/trans.html`. Licensed under Public Domain via Commons.

36. In the distant past, pebbles or scratches or other marks were used to record quantities; see Chapter 2 [Rudman, 2007]. This could be done without any number concept: Each pebble or mark simply

corresponded to one item. For example, pebbles would represent the sheep of a herd leaving the village in the morning. When the herd returned from the pasture in the evening, the pebbles were matched with the sheep. When all pebbles had been used up, the entire herd had been accounted for. To simplify the process, names were made up for groups of pebbles or scratch marks. This led to counting and the natural numbers.

37. The philosophical question "Is mathematics discovered or invented?" has been debated for centuries. We have argued for invention for a number of years. Our most recent (final?) investigation of the question in [Truemper, 2022] explains the reason the debate has been raging for centuries and may well continue unabated in the future.

38. See Wikipedia "Leopold Kronecker."

39. See Wikipedia "Finitism."

40. Source: https://en.wikipedia.org/wiki/Leopold_Kronecker# /media/File:Leopold_Kronecker_1865.jpg. "Leopold Kronecker 1865" by Unknown. Licensed under Public Domain via Commons.

41. Source: https://en.wikipedia.org/wiki/Giuseppe_Peano#/ media/File:Giuseppe_Peano.jpg. "Giuseppe Peano" by Unknown - School of Mathematics and Statistics, University of St Andrews, Scotland [1]. Licensed under Public Domain via Commons.

42. See Wikipedia "Natural number."

43. There are a total of five axioms, listed below with comments in parentheses. The axioms use a constant symbol 0 and a unary function symbol S. ($S(n)$ is the successor $n + 1$ of a natural number n.)

1. 0 is a natural number. (The constant 0 is assumed to be a natural number.)
2. For every natural number n, $S(n)$ is a natural number.
3. For all natural numbers m and n, $m = n$ if and only if $S(m) = S(n)$. (S is an injection.)
4. For every natural number n, $S(n) = 0$ is false. (There is no natural number whose successor is 0.)
5. Let K be a set such that the following holds: 0 is in K, and for every natural number n, if n is in K, then $S(n)$ is in K. Then K contains every natural number. (Induction axiom.)

See Wikipedia "Giuseppe Peano" and "Peano Axioms" for further details. Note that Peano considers 0 to be a natural number as a matter of mathematical convenience. The number has a complicated history, see Wikipedia "0."

44. See Wikipedia "Dedekind cut."

45. Peano considers 0 to be a natural number as a matter of mathematical convenience.

46. An important equivalent statement of the induction axiom assumes that the claim to be proved holds for $n = 0$ but fails for some larger n. The axiom then states that there is a smallest $n > 0$ where the claim fails.

Chapter 3 Notation

47. Source: https://en.wikipedia.org/wiki/Differential_calculus#/media/File:Tangent_to_a_curve.svg. "Tangent to a curve" by Jacj. Later versions were uploaded by Oleg Alexandrov at en.wikipedia. - Transferred from en.wikipedia to Commons. Licensed under Public Domain via Commons.

48. Source: https://en.wikipedia.org/wiki/Integral#/media/File:Integral_example.svg. "Integral example" by I, KSmrq. Licensed under CC BY-SA 3.0 via Commons.

49. [Cajori, 1919b].

50. Source: https://en.wikipedia.org/wiki/Isaac_Newton#/media/File:GodfreyKneller-IsaacNewton-1689.jpg. "IsaacNewton-1689" by Godfrey Kneller, 1689. Licensed under Public Domain via Commons.

51. Source: https://en.wikipedia.org/wiki/Philosophi%C3%A6_Naturalis_Principia_Mathematica#/media/File:Prinicipia-title.png. "Principia-title" originally uploaded by Zhaladshar at Wikisource - Transferred from en.wikisource to Commons.

52. [Cajori, 1919b] includes relevant excerpts of the Principia, including an English translation of the material taken from the third edition.

53. Source: https://en.wikipedia.org/wiki/Gottfried_Wilhelm_Leibniz#/media/File:Gottfried_Wilhelm_von_Leibniz.jpg. "Gottfried Wilhelm von Leibniz" by Christoph Bernhard Francke -

/gbrown/philosophers/leibniz/BritannicaPages/Leibniz/Leibniz Gif.html. Licensed under Public Domain via Commons.

54. Source: The Lilly Library, Indiana University, Bloomington, Indiana, kindly granted use of the photo.

55. For a detailed analysis, see pp. 85–95 [Grötschel et al., 2016]. Wikipedia "Leibniz–Newton calculus controversy" has a summary.

56. For example, the slope of the function f is denoted by $\dot{f}$, the slope of the slope is $\ddot{f}$, and the integral is denoted by $\bar{f}$.

57. In modern mathematics, slope is denoted by $\frac{df}{dx}$, and the slope of the slope by $\frac{d^2 f}{dx^2}$. For the evaluation of the area under the function $f(x)$ over the range $x = a$ to $x = b$, the expression is $\int_a^b f(x)dx$.

58. Source: https://en.wikipedia.org/wiki/Guillaume_de_l'H ôpital#/media/File:Guillaume_de_l'Hôpital.jpg. "Guillaume de
l'Hôpital." Licensed under Public Domain via Commons.

59. The title of the book is "Analyse des Infiniment Petits pour l'Intelligence des Lignes Courbes." Recent research results confirm that the book was largely based on the work of Johann Bernoulli. In the introduction of the book, l'Hôpital acknowledges this, writing "I must own myself very much obliged to the labours of Messieurs Bernoulli, but particularly to those of the present Professor at Groeningen [Johann Bernoulli], as having made free use of their Discoveries as well as those of Mr. Leibnitz: So that whatever they please to claim as their own I frankly return them." The translation, slightly modified, is taken from http://faculty.wlc.edu/buelow/calc/nt4-5.html. See also Wikipedia "Analyse des Infiniment Petits pour l'Intelligence des Lignes Courbes."

60. The proof that the slope of the function at a maximum or minimum must be 0, follows almost directly from the definition of slope: If at such a point the slope was positive, then a small increase of the variable would increase the function value, and a decrease of the variable would decrease the function value. If the slope was negative, the effects of the two changes of the variable would be reversed.

61. For details, see Wikipedia "Brachistochrone curve," including the mathematical description of the curve.

62. Source: https://en.wikipedia.org/wiki/File:Justus_Sus
termans_-_Portrait_of_Galileo_Galilei,_1636.jpg. "Justus
Suster-
mans - Portrait of Galileo Galilei, 1636" by Justus Sustermans. Licensed under Public Domain via Commons.

63. Source: https://en.wikipedia.org/wiki/Johann_Bernoulli
#/media/File:Johann_Bernoulli2.jpg. "Johann Bernoulli2" by Jo-
hann Rudolf Huber. Licensed under Public Domain via Commons.

64. See Wikipedia "Brachistochrone curve."

65. See Wikipedia "Brachistochrone curve."

66. See Wikipedia "Calculus of variations."

67. Source: http://www.taralaya.org/science-park/BRACHIST
OCHRONE_clip_image002.jpg. "Brachistochrone demonstration"
photo copyright Jawaharlal Nehru Planetarium, Bangalore, India.
The Planetarium kindly granted use of the photo.

68. Source: "Richard E. Bellman" copyright RAND Corporation,
which kindly permitted use of the photo.

69. Proof that the segment New York – St. Louis is the shortest
route: If that is not the case, then there is a shorter route connecting the two cities. But then that shorter route plus the St. Louis –
Los Angeles segment constitute a shorter route New York – Los
Angeles.

70. See Wikipedia "Dynamic Programming."

71. [Dreyfus, 1965].

72. Search the internet using "NASA dynamic programming" for
a long list of papers on control problems of space exploration that
are solved with dynamic programming.

73. Chapter 4, 5 [Rudman, 2007].

74. See Wikipedia "Mechanical computer."

75. The text was later translated into Arabic, Latin, and English. It
became one of the most influential books in the history of mathematics; see [Heath, 1910].

76. Paragraph 101 [Cajori, 1928].

77. Source: https://en.wikipedia.org/wiki/Arithmetica#/med ia/File:Diophantus-cover.jpg. "Arithmetica by Diophantus." Licensed under Public Domain via Commons.

78. Folio 249 verso [Stifel, 1544]. See also Wikipedia "Michael Stifel."

79. Source: https://en.wikipedia.org/wiki/Michael_Stifel#/ media/File:Michael_Stifel.jpeg. "Michael Stifel" by Unknown. Licensed under Public Domain via Commons.

80. For example, if we want to compute $\frac{1}{8} \cdot 4$, we look up the exponents of $\frac{1}{8}$ and 4, which are -3 and 2, add them together and get -1, and finally look up for the latter exponent the corresponding number, which is $\frac{1}{2}$. Thus, $\frac{1}{8} \cdot 4 = \frac{1}{2}$.

81. Division: $2^m / 2^n = 2^{m-n}$. Powers: $(2^m)^n = 2^{m \cdot n}$. Root: $\sqrt[n]{2^m} = 2^{m/n}$.

82. See de.wikipedia "Jost Bürgi."

83. Source: https://en.wikipedia.org/wiki/Jost_B%C3%BCrgi# /media/File:Jost_B%C3%BCrgi_Portr%C3%A4t.jpg. "Jost Bürgi Porträt" by User Dvoigt on de.wikipedia. Licensed under Public Domain via Commons.

84. Source: https://en.wikipedia.org/wiki/Jost_B%C3%BCrgi#/m edia/File:JostBurgi-MechanisedCelestialGlobe1594.jpg. "Jost Buergi-Mechanised Celestial Glob2" photo by Horology - Own work. Licensed under CC BY-SA 3.0 via Commons.

85. Bürgi had an exponent notation for variables that is connected with Stifel's exponent. He wrote a Roman numeral above a constant as the exponent of a variable; the constant was the coefficient of the variable. For example, the term $\overset{vi}{4}$ has the Roman numeral $vi = 6$ above the number 4; the expression denotes $4x^6$ in modern notation; see paragraph 296 [Cajori, 1928].

86. Bürgi's choice of $B = 1.0001$ was really clever since successive multiplication by B amounted to the following: Take the number already on hand for a given n value, write it again below while shifting it four positions to the right, and add the two numbers. Voilà, there is the number for $n + 1$. Thus, the numbers for the exponents $n = 0, \dots, 23027$ could readily be computed, starting with $n = 0$.
Some additional steps were needed to avoid build-up of numeri-

cal errors; see p. 15 [Waldvogel, 2014]. But even with such checks and corrections, the entire construction process of the tables was extraordinarily efficient. The estimate of several months of manual effort is included on p. 15 [Waldvogel, 2014].

87. p. 6 [Waldvogel, 2014].

88. [Bürgi, 1620].

89. The manual is available on pp. 26–36 [Gieswald, 1856].

90. Source: Toggenburger Museum, Lichtensteig, Switzerland. The museum kindly has granted permission to use the photo.

91. See Wikipedia "John Napier."

92. p. 188, 189 [Cajori, 1919a] explains Napier's line of thought as follows. Imagine two lines, the first one with two marks A and B, and second one with just one mark D. A point moves from A to B, and a second point from D in one of the two possible directions. The points start at the same time and with the same initial velocity. The first point immediately starts to slow down: At any intermediate point between A and B, say C, the initial velocity is reduced by the factor (distance C to B)/(distance A to B). In contrast, the second point never changes its velocity.
Suppose the first point has arrived at C as just stated, and the second one is at a point F at the same time. Napier calls the distance from D to F the *logarithm* of the distance from C to B.
The modern-day logarithm function is 0 when its argument is 1; in Napier's tables, the distance from A to B is 10^7, so 0 is Napier's logarithm of 10^7. Thus, his approach didn't suggest the concept of *base*; indeed, that concept seemingly is inapplicable to his method. But when movements of the two points are analyzed, it becomes clear that Napier's logarithm for a number x is equal to $10^7 (\ln 10^7 - \ln x)$, where ln denotes the *natural logarithm*, which has as base the Euler number $e = 2.7182\ldots$.

93. Source: `https://en.wikipedia.org/wiki/John_Napier#/med ia/File:John_Napier.jpg`. "John Napier" by Unknown - scanned from `http://www-history.mcs.st-and.ac.uk/history/PictDispl ay/Napier.html`. Licensed under Public Domain via Commons.

94. Source: `https://en.wikipedia.org/wiki/John_Napier#/media /File:Logarithms_book_Napier.jpg`. "Logarithms book Napier" by Unknown - Napier, Mark (1834), William Blackwood. Licensed under Public Domain via Commons.

95. See Wikipedia "Henry Briggs (mathematician)."

96. [Gieswald, 1856] contains a detailed discussion. See also the arguments in [Waldvogel, 2014] and Wikipedia "History of logarithms."
The book *The Daring Invention of Logarithm Tables* ([Truemper, 2020]) describes the efforts of Bürgi, Napier, and Briggs in detail and the subsequent development of the slide rule, circular slide rule, and slide cylinder. The book is also available in German: *Die wagemutige Erfindung der Logarithmentafeln* ([Truemper, 2024]).

97. See Wikipedia "René Descartes."

98. [Descartes, 1637].

99. Source: `https://en.wikipedia.org/wiki/Ren%C3%A9_Desc artes#/media/File:Frans_Hals_-_Portret_van_Ren%C3%A9_D escartes.jpg`. "Frans Hals - Portret van René Descartes" after Frans Hals (1582/1583–1666) - André Hatala [e.a.] (1997) De eeuw van Rembrandt, Bruxelles: Cédit communal de Belgique, ISBN 2-908388-32-4. Licensed under Public Domain via Commons.

100. Source: `https://en.wikipedia.org/wiki/Discourse_on_the _Method#/media/File:Descartes_Discours_de_la_Methode.jpg`. "Discours de la Méthode" by Unknown. Licensed under Public Domain via Commons.

101. [Rudman, 2007].

102. Archimedes by Domenico Fetti, 1620. Source: `https://en.wik ipedia.org/wiki/File:Domenico-Fetti_Archimedes_1620.jpg`. Painting of Alte Meister Museum, Dresden, Germany; see `http: //archimedes2.mpiwg-berlin.mpg.de/archimedes_templates/p opup.htm`. Public Domain under US copyright code PD-old-100.

103. "Archimedes parabola with triangle" by K. Truemper, released into Public Domain under Creative Commons CC0.

104. Source: `https://en.wikipedia.org/wiki/Archimedes#/me dia/File:Esfera_Arqu%C3%ADmedes.jpg`. "Esfera Arquímedes" by Andertxuman - Own work. Licensed under Public Domain via Commons.

105. [Netz and Noel, 2007] contains historical and mathematical details; see also Wikipedia "Archimedes."

106. Paragraph 340 [Cajori, 1928].

107. Paragraph 298 [Cajori, 1928].

108. See Wikipedia *"La Géométrie"* and "Cartesian coordinate system."

109. Source: `https://en.wikipedia.org/wiki/Cartesian_coordi nate_system#/media/File:Cartesian-coordinate-system.svg`. "Cartesian coordinate system." By K. Bolino - Made by K. Bolino (Kbolino), based upon earlier versions. Licensed under Public Domain via Commons.

110. Source: `https://en.wikipedia.org/wiki/Cartesian_coordi nate_system#/media/File:Cartesian-coordinate-system-with -circle.svg`. "Cartesian-coordinate-system-with-circle" by 345Kai. Licensed under CC BY-SA 3.0 via Commons.

111. See Wikipedia "History of the function concept."

112. Source: `https://en.wikipedia.org/wiki/Function_(ma thematics)#/media/File:Function_machine2.svg`. "Function machine2" by Wvbailey (talk) - Own work (Original text: I created this work entirely by myself.). Licensed under Public Domain via Commons.

113. Source: `https://en.wikipedia.org/wiki/Graph_of_a_funct ion#/media/File:Three-dimensional_graph.png`. "Three-dimensional graph" by dino (talk) - Own work. Licensed under CC BY-SA 3.0 via Commons. The depicted function is $f(x, y) = sin(x^2) \cdot cos(y^2)$.

114. A function is continuous if changes of $f(x)$ can always be made arbitrarily small by suitably restricting changes of x.

115. See Wikipedia "Alan Turing" and "Halting problem."

116. Source: `https://en.wikipedia.org/wiki/Alan_Turing#/m edia/File:Alan_Turing_Aged_16.jpg` "Alan Turing Aged 16" by Unknown - `http://www.turingarchive.org/viewer/?id=521&tit le=4`. Licensed under Public Domain via Commons.

117. The natural number representing a computer program is determined as follows. Each instruction of the computer program is coded by a natural number. The string of these natural numbers, viewed as one large natural number, then represents the entire computer program.

118. The results claimed for the number of functions follow directly

from Cantor's work discussed in Chapter 4. Essentially, the number of functions with the natural numbers as input and 0 or 1 as output is the same as the number of real numbers, and the number of functions with the real numbers as input and 0 or 1 as output is equal to the number of subsets of the set of real numbers. In the notation of Chapter 4, the first number is $2^{\aleph_0}$, which is infinitely larger than the number of natural numbers $\aleph_0$. The second number is $2^{2^{\aleph_0}}$, which is infinitely larger than $2^{\aleph_0}$.

119. See Wikipedia "Function (mathematics)."

120. Source: https://en.wikipedia.org/wiki/Bernhard_Riemann#/media/File:Georg_Friedrich_Bernhard_Riemann.jpeg. "Georg Friedrich Bernhard Riemann" by https://de.wikipedia.org/wiki/Bernhard_Riemann#/media/File:BernhardRiemannAWeger.jpg of de.wikipedia. Licensed under Public Domain via Commons.

121. Source: https://commons.m.wikimedia.org/w/index.php?search=Henri+Lebesgue+&fulltext=search#/media/File%3ALebesgueH.gif. By Unknown. Public Domain since copyright has expired.

122. In mathematical notation, the width of a slice is dx, the height of a slice is $f(x)$, and the estimate of the area of the slice is $f(x)dx$. When dx becomes ever smaller, then under suitable assumptions the sum of these areas, denoted by $\int f(x)dx$, converges to the desired value.

123. For a vertical strip of width dx, estimation of the strip's area by $f(x)dx$ is successful only if the function does not jump within the interval. Hence, any function that jumps within the interval, no matter how narrow, cannot be processed.

124. Source: https://en.wikipedia.org/wiki/Lebesgue_integration#/media/File:RandLintegrals.png. "RandLintegrals" included in en.wikipedia. Licensed under CC BY-SA 3.0 via Commons.

125. In mathematical notation, the height of a horizontal slice is dt, and the length of the slice is the sum of the lengths of the segments of the x-axis for which the function value is greater than t. Using a function μ that can add up the length of these line segments, the length of the slice is computed as $f^*(t) = \mu(\{x \mid f(x) > t\})$. The Lebesgue integral of f is then $\int f\,d\mu = \int_0^\infty f^*(t)\,dt$ where the integral on the right is a Riemann integral.

126. See Wikipedia "Lebesgue integration."

127. See Wikipedia "Lebesgue integration."

128. The Lebesgue integral can be computed since the total length of the line segments for which $f(x) > 0$ can be proved to be 0.

Chapter 4 Infinity

129. Leonardo Donato rebelled against the power of Pope Paul V in a struggle whose outcome is best characterized as a draw; see Wikipedia "Leonardo Donato."

130. Source: https://commons.wikimedia.org/wiki/File:Galileo_Donato.jpg#/media/File:Galileo_Donato.jpg. "Galileo Donato" by H. J. Detouche - http://www.astro.unipd.it/insap6/mainPage.html. Licensed under Public Domain via Commons.

131. Source: https://en.wikipedia.org/wiki/Nicolaus_Copernicus#/media/File:Nikolaus_Kopernikus.jpg. "Nikolaus Kopernikus" by Unknown - http://www.frombork.art.pl/Ang10.htm. Licensed under Public Domain via Commons.

132. pp. 83–85 [Alexander, 2014].

133. p. 138 [Alexander, 2014]. See also Wikipedia "Galileo Galilei."

134. [Alexander, 2014] describes in detail the new ideas and the forces trying to suppress them.

135. See Wikipedia "Bonaventura Cavalieri."

136. Source: https://en.wikipedia.org/wiki/Bonaventura_Cavalieri#/media/File:Bonaventura_Cavalieri.jpeg. "Bonaventura Cavalieri." Licensed under Public Domain via Common.

137. Source: https://archive.org/. Search for "Bonaventura Cavalieri." Public Domain .

138. See Wikipedia "Cavalieri's principle."

139. *Geometria indivisibilibus continuorum nova quadam ratione* is available at https://archive.org/. Search for "Bonaventura Cavalieri."

140. See Wikipedia "Bonaventura Cavalieri."

141. See Chapter 3.

142. See Wikipedia "Evangelista Torricelli."

143. Source: https://en.wikipedia.org/wiki/Evangelista_Torr
icelli#/media/File:Evangelista_Torricelli_by_Lorenzo_Li
ppi_%28circa_1647,_Galleria_Silvano_Lodi_%26_Due%29.jpg.
"Evangelista Torricelli" by Lorenzo Lippi, ca. 1647. Galleria Silvano
Lodi & Due. Licensed under Public Domain via Commons.

144. Source: Joern Koblitz of Milestones of Science Books, Bremen,
Germany, kindly granted permission to use photo of title page of
Torricelli's "Opera Geometrica," all rights reserved.

145. Source: "Rectangle and Parallelogram" by K. Truemper, released into Public Domain under Creative Commons CC0.

146. Source: "Hyperbola" by K. Truemper, released into Public Domain under Creative Commons CC0.

147. Source: https://en.wikipedia.org/wiki/Gabriel%27s_Horn
#/media/File:GabrielHorn.png. "Gabriel's Horn" by RokerHRO -
Own work. Licensed under Public Domain via Commons.

148. See Wikipedia "Gabriel's Horn" for details of the following
formulas. The volume V and surface area A of the trumpet trun-
cated at some point $a > 1$ are given by $V = \pi \int_1^a \left(\frac{1}{x}\right)^2 dx =$
$\pi \left(1 - \frac{1}{a}\right)$ and $A = 2\pi \int_1^a \frac{1}{x}\sqrt{1 + \left(\frac{-1}{x^2}\right)^2} dx > 2\pi \int_1^a \frac{1}{x} dx = 2\pi \ln a$.
When a goes to ∞, we get $\lim_{a\to\infty} V =$
$\lim_{a\to\infty} \pi \left(1 - \frac{1}{a}\right) = \pi$ and $\lim_{a\to\infty} A \geq \lim_{a\to\infty} 2\pi \ln a = \infty$.

149. It's easy for us today to exhibit the flaw in the paint argument.
At the time, the finite volume and infinite surface of the trumpet
seemingly were a paradox, and Torricelli attempted several proofs
to show that the surface was finite. See Wikipedia "Evangelista Torricelli."

150. See Chapters 3–5 of [Alexander, 2014] for details about the
fight of the Jesuits against the idea of indivisibles and the devastating impact of their victory.

151. Source: https://en.wikipedia.org/wiki/John_Wallis#/med
ia/File:John_Wallis_by_Sir_Godfrey_Kneller,_Bt.jpg. Public
Domain under US copyright code PD-old-100 .

152. Source: go to https://archive.org/ and search for "john
wallis de sectionibus conicis." Licensed under Public Domain Mark

1.0 Creative Commons.

153. See Wikipedia "John Wallis."

154. Here is an example of Wallis's method. In Proposition 3 of *De sectionibus conicis*, he computes the area of triangular figures as follows. He first slices the area of such a figure horizontally into an infinite number of strips. Let the base line of the figure have length B and the height of the figure be A. Since the number of strips is equal to ∞, each strip has height $\frac{A}{\infty}$. Since the length of the strips goes from B to 0 evenly as one moves from the baseline to the top point of the figure, the average strip length is $\frac{B}{2}$. Since there are an infinite number of strips, the total length of all strips is ∞ times the average strip length, that is, $\infty \cdot \frac{B}{2}$. The total area is the product of strip height times total strip length, so Area $= \frac{A}{\infty} \cdot \infty \cdot \frac{B}{2}$. The two instances of ∞ cancel out, and thus Area $= \frac{A \cdot B}{2}$.

155. See Wikipedia "John Wallis."

156. See Wikipedia "Wallis product."

157. At the time of Newton and Leibniz, the function concept was yet to be introduced; see Chapter 3. Instead, Newton and Leibniz treated formulas. For clarity, we use the function concept.

158. See Chapter 3.

159. We use d for a small change of x to discuss both Newton's and Leibniz's methods. Newton actually employed o for a small quantity, while Leibniz used dx.

160. Source: "Differential" by K. Truemper, released into Public Domain under Creative Commons CC0.

161. The description of Leibniz's process in terms of x, $f(x)$ and d is necessarily imprecise since the function concept was unknown at that time. We use discussion of the $f(x) = x^2$ case by Bernoulli as cited on p. 28 [Bos, 1974], which roughly matches the description used here.

162. p. 9 [Cajori, 1919b] contains Newton's arguments:
"It is objected, that there is no ultimate proportion of evanescent quantities [here, $f(x + d) - f(x)$ and d are the evanescent quantities]; because the proportion, before the quantities have vanished, is not ultimate; and when they have vanished, is none. But, by the same argument, it might as well be maintained, that there is no

ultimate velocity of a body arriving at a certain place, when its motion is ended: because the velocity, before the body arrives at the
place, is not its ultimate velocity; when it has arrived, is none.
"But the answer is easy: for the ultimate velocity is meant that, with
which the body is moved, neither before it arrives at its last place,
when the motion ceases, nor after; but at the very instant when it
arrives; that is, the very velocity with which the body arrives at its
last place, when the motion ceases.
"And, in like manner, by the ultimate ratio of evanescent quantities
is to be understood the ratio of the quantities, not before they vanish, nor after, but that with which they vanish."
Why is this analogy wrong? Say, the body is an arrow. When it hits
a target, the velocity of the arrow does not instantaneously drop to
0. If that were so, the arrow would hit the target with infinite force.
Hence the image of an arrow hitting a target does not supply the
desired conclusion.

163. See the overview of Leibniz's method in [Bos, 1974].

164. Source: https://en.wikipedia.org/wiki/Bernard_Bolzano#
/media/File:Bernard_Bolzano.jpg. "Bernard Bolzano." Licensed
under Public Domain via Commons.

165. Source: https://en.wikipedia.org/wiki/Augustin-Lou
is_Cauchy#/media/File:Augustin-Louis_Cauchy_1901.jpg.
"Augustin-
Louis Cauchy 1901" by Library of Congress Prints and Photographs
Division. From an illustration in: Das neunzehnte Jahrhundert in
Bildnissen / Karl Werckmeister, ed. Berlin: Kunstverlag der Pho-
tographischen Gesellschaft, 1901, vol. V, no. 581. Licensed under
Public Domain via Commons.

166. The technical definition of convergence of a sequence $S = s_1$,
$s_2, s_3, \ldots$ to a limit L is as follows: For any value $\delta > 0$, there is an
index n such that, for all $i > n$, the terms s_i satisfy $|s_i - L| < \delta$.

167. The stated definition of continuity isn't practically useful: To
verify continuity just at one point c, we must consider all sequences
$x_1, x_2, x_3, \ldots$ that converge to c, a potentially difficult task.
Weierstrass created an alternate and eminently useful definition using ϵ and δ: For any tolerance $\epsilon > 0$—no matter how small—there
is a value $\delta > 0$ such that, whenever x satisfies $c - \delta < x < c + \delta$,
then $f(c) - \epsilon < f(x) < f(c) + \epsilon$.
The condition can be stated in terms of the drawing below as follows: Given any $\epsilon > 0$—no matter how small—let the horizontal

borders of the horizontal strip be defined by $f(c) - \epsilon$ and $f(c) + \epsilon$ as shown. Then there must be $\delta > 0$ defining the vertical borders of the vertical strip by $c - \delta$ and $c + \delta$, such that the portion of the function $f(x)$ falling within the vertical strip, here indicated by a heavy segment, is also contained in the horizontal strip.

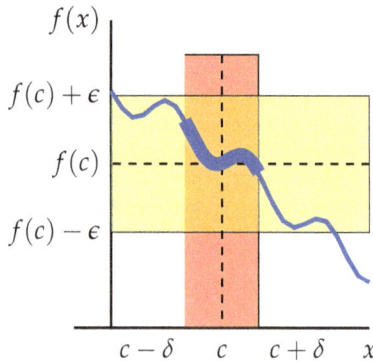

(Source: "Delta/Epsilon Continuity Definition" by K. Truemper, released into Public Domain under Creative Commons CC0.)

168. Source: https://en.wikipedia.org/wiki/Karl_Weierstra ss#/media/File:Karl_Weierstrass.jpg. "Karl Weierstrass." Licensed under Public Domain via Commons.

169. Source: "Continuous function" by K. Truemper, released into Public Domain under Creative Commons CC0.

170. Source: "Function with jump" by K. Truemper, released into Public Domain under Creative Commons CC0.

171. See Wikipedia "Limit (mathematics)."

172. Source: "Extended Function" by K. Truemper, released into Public Domain under Creative Commons CC0.

173. See Wikipedia "Mathematical analysis."

174. For discussion of the case of a general function $f(x)$, let's use Leibniz's dx instead of d and the limit notation. We then have the derivative $\frac{df}{dx} = \lim_{dx \to 0} \frac{f(x+dx)-f(x)}{dx}$. The limit notation emphasizes that the slope of $f(x)$ at a point x is computed by finding a continuous extension of the function $\frac{f(x+dx)-f(x)}{dx}$ where x is an arbitrary point, dx is the variable, and the extension is to be found for $dx = 0$. How do we know that this extension will always exist? Well, we do not. For example, if the function jumps at a point x,

then $f(x + dx) - f(x)$ jumps at $dx = 0$, and $\frac{f(x+dx)-f(x))}{dx}$ has no continuous extension at $dx = 0$.

175. For a long time, it was believed that slope could always be computed for continuous functions except for some exceptional points. For the Weierstrass function, the desired continuous extension of $\frac{f(x+dx)-f(x)}{dx}$ does not exist for *any* point x. In fact, it is now known that virtually *all* continuous functions exhibit this behavior! See Wikipedia "Weierstrass function" for details.
As an aside, suppose we declare that the continuous functions employed in the sciences and engineering, which typically are differentiable except maybe at some special points, are *normal*, and that nondifferentiable continuous functions like the Weierstrass function are *pathological* cases. Then according to the cited result, virtually all continuous functions are pathological, a strange conclusion. It is one more indication that mathematics is not part of nature.

176. Galileo established a paradox: He wrote below the natural numbers 1, 2, 3, 4 ... their square values 1, 4, 9, 16 ... and thus concluded that there were just as many square values as there were natural numbers. Yet the second list did not contain all natural numbers, and hence should be smaller. He concluded that ideas such as *less*, *equal*, and *greater* applied to finite quantities but not infinite ones. See Wikipedia "Galileo's paradox."

177. See Wikipedia "Finitism."

178. See Wikipedia "Actual infinity."

179. Cantor compares the set N of natural numbers with any other infinite set T as follows.

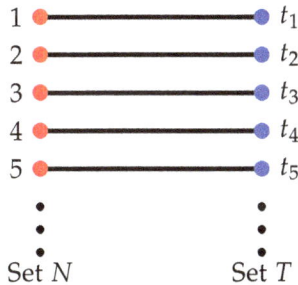

(Source: "Countable set" by K. Truemper, released into Public Domain under Creative Commons CC0.)
First, he checks if each element of N can be assigned to a different element of T as shown above, where the elements 1, 2, 3, ... on the

left are in N and the elements t_1, t_2, t_3, ... on the right in T. When this is the case, he declares that the cardinality of T is greater than or equal to that of N. In shorthand notation, $|T| \geq |N|$.

Then he checks if each element of T can be assigned to a different element of N. When that is the case, he has $|T| \leq |N|$. When both cases apply, he declares that N and T have the same cardinality, denoted by $|N| = |T|$. The same approach is used to compare the cardinality of arbitrary infinite sets.

180. It is by no means self-evident that the set N of natural numbers has the smallest cardinality of all infinite sets. But when the *axiom of countable choice* is added to Zermelo-Fraenkel set theory—more on this in Chapter 6—then it can be proved that N indeed represents the smallest infinity. Indeed, it can be shown that any infinite set T contains a countable infinite subset, that is, a subset of the form $\{t_1, t_2, t_3, \dots\}$. Hence the elements of N can be assigned to the elements of the subset, and $|N| \leq |T|$ has been proved. For details, see Wikipedia "Axiom of countable choice."

181. Proof: Let $R = \{r_1, r_2, r_3 \dots\}$ and $S = \{s_1, s_2, s_3 \dots\}$. Create T from R by replacing each element r_i by the elements of S, notationally modified so that they are unique, say denoted by s_1^i, s_2^i, s_3^i, ... There are just as many s_j^i in T as there are pairs (r_i, s_j). For simplicity, assume that each r_i and s_j is a natural number. By the fundamental theorem of arithmetic—see Chapter 2—the prime numbers 2 and 3 can be used to uniquely represent each pair (r_i, s_j) by the natural number $2^{r_i} \cdot 3^{s_j}$. The set of the latter numbers is a subset of the natural numbers and thus countable, and so is T.

182. Each rational number is a ratio of two integers m and n. Thus, by substitution, the ratios can be counted. For the algebraic numbers, the polynomials with integer coefficients can be counted, and each of them has a finite number of solutions. By substitution, the solutions can be counted.

183. Proof: Use the fact that the set of algebraic numbers is countable, then use repeated substitution to handle n-dimensional space.

184. Consider the real numbers between 0 and 1, represented in binary notation. An example is $0.1001011\dots = 2^{-1} + 2^{-4} + 2^{-6} + 2^{-7}\dots = 0.58\dots$ in decimal notation. It suffices to show that there are more real numbers than natural ones, since it has already been shown that the number of natural numbers matches that of the rational and algebraic numbers.

The result is established by showing that, when one tries to assign the real numbers to the natural numbers, then not all of them can be accommodated. Assume such an assignment is possible, say where for $i = 1, 2, 3, \ldots$, the real number r_i is assigned to i. Since all digits of each r_i are either 0 or 1, define a new number r^* whose ith digit is the opposite of the ith digit of r_i. Thus, r^* is different from all r_i and thus has not been assigned, a contradiction. For further details about the diagonal argument, see Wikipedia "Cantor's diagonal argument."

185. Consider the binary representation of the real numbers r between 0 and 1. Take the example $r = 0.1001011\ldots$, and look at the 1s of r. They are in positions $1, 4, 6, 7, \ldots$ These indices correspond to the subset $\{1, 4, 6, 7, \ldots\}$ of the set N natural numbers, so the subset is a unique representation of r. Conversely, the subset uniquely defines the 1s of r, and r is a unique representation of the subset. Thus, the subsets of N are in one-to-one correspondence with the real numbers r between 0 and 1.
The subsets of a given set S form a set called the *power set* of S. If S is finite and has n elements, then the power set of S has 2^n elements.
Cantor extends this notation to infinite sets. Since N has cardinality $\aleph_0$, he declares that the power set of N has cardinality $2^{\aleph_0}$. Due to the above derived correspondence of the power set of N and the set of real numbers between 0 and 1, the latter set has cardinality $2^{\aleph_0}$ as well. By a trivial substitution step, the same conclusion applies to the set of all real numbers.

186. The expanded conclusion for the algebraic numbers relies on the fact that the proof of the original statement involving the rational numbers uses only the fact these numbers are countable. Now the algebraic numbers are countable as well, and the result follows.

187. The natural numbers allow an ordering of finite sets. For example, suppose we have a set with six elements. We can label them $a_1, a_2, \ldots, a_6$ and thus order them.
Cantor's concept of *ordinal numbers*, for short *ordinals*, extends this idea to infinite sets. Given an infinite set, ordinal numbers can be used to label the elements and thus obtain an ordering.
Ordinals are different from the *cardinal numbers*, for short *cardinals*, such as $\aleph_0$: The latter concept just measures the number of elements in a set and is not concerned with any ordering of the elements. Cantor introduced ordinals so that he could define and manipulate infinite sequences.

The finite ordinals, as well as the finite cardinals, are just the natural numbers, including the 0. So we have 0, 1, 2, ... The smallest infinite ordinal is labeled ω and is associated with the cardinal $\aleph_0$. But for the many possible sets with cardinality $\aleph_0$, a huge number of orderings are possible. This is reflected in a correspondingly vast number of ordinals. An appealing intuitive definition of these and subsequent ordinals is given in Wikipedia "Ordinal number," as follows.

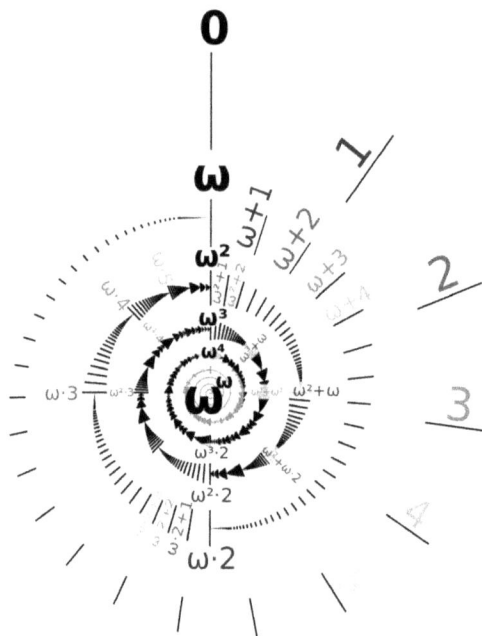

(Source: https://commons.wikimedia.org/wiki/File:Omega-exp-omega-labeled.svg. "Omega-exp-omega-labeled" by Pop-up casket (talk); original by User:Fool - Own work. Licensed under CC0 via Commons.)

Start with the natural number 0 of the above spiral, and proceed clockwise. The next ordinal numbers are 1, 2, ... After all natural numbers comes the first infinite ordinal, ω, and after that $\omega + 1$, $\omega + 2$, $\omega + 3$, ... Ignore the meaning of the "+" sign, and consider these terms just to be labels.

Next comes $\omega + \omega$, denoted by $\omega \cdot 2$, then $\omega \cdot 2 + 1$, $\omega \cdot 2 + 2, \ldots$, then $\omega \cdot 3$, and later $\omega \cdot 4$. Now the ordinals $\omega \cdot m + n$, where m and n are natural numbers, formed by this process are followed by ω^2. Going on, we encounter ω^3, $\omega^4, \ldots$, then ω^ω.

At that point, the spiral stops, but the list of ordinals actually goes on. Next are ω^{ω^2}, ω^{ω^3}, ..., and much later an ordinal called ϵ^0. And this constitutes a list of relatively small, countable ordinals! Indeed, we can continue this construction indefinitely. The smallest uncountable ordinal is the set of all countable ordinals, expressed as ω_1. Cantor denotes the cardinality of that set by $\aleph_1$.

188. See Wikipedia "Georg Cantor."

189. Source: `https://en.wikipedia.org/wiki/G%C3%B6sta_Mitta g-Leffler#/media/File:Magnus_Goesta_Mittag-Leffler_1.jpg`. "Magnus Goesta Mittag-Leffler 1" by Unknown. Licensed under Public Domain via Commons.

190. The letters are summarized in [Schoenflies, 1927], written for the celebration of Mittag-Leffler's 80th birthday. The paper brings out Mittag-Leffler's central role in the support of Cantor's work. As an aside, Mittag-Leffler was a strong advocate for women. For example, because of his efforts, Sofia Kovalevskaya, the first major Russian female mathematician, became the first woman anywhere in world to hold the position of full professor at a university; see Wikipedia "Mittag-Leffler" and "Sofia Kovalevskaya."

Chapter 5 Six Problems of Antiquity

191. See Chapter 3.

192. The latter effort culminated in Euler's famous equation $e^{i\pi} + 1 = 0$, which brings together the transcendental e and π, the imaginary i, and the fundamental 0 and 1. The physicist Richard Feynman called the equation "our jewel" and "the most remarkable formula in mathematics." See Wikipedia "Euler's formula."

193. See Chapter 4.

194. See Wikipedia "Compass equivalence theorem" for the simple construction.

195. Source: "Non-collapsing compass" by K. Truemper, released into Public Domain under Creative Commons CC0.

196. "Triangle, square, and pentagon" by K. Truemper, released into Public Domain under Creative Commons CC0.

197. Source: "Parabola" by K. Truemper, released into Public Domain under Creative Commons CC0.

198. Source: "Trisection" by K. Truemper, released into Public Domain under Creative Commons CC0.

199. Source: "Double cube" by K. Truemper, released into Public Domain under Creative Commons CC0.

200. Source: "Square circle" by K. Truemper, released into Public Domain under Creative Commons CC0.

201. There are a number of equivalent formulations of the fifth axiom; see Wikipedia "Euclidean geometry." We list Playfair's version; see Wikipedia "Playfair's axiom." Note that the first four axioms imply that there is at least one parallel line. Thus, the condition of Playfair's axiom that there is *at most* one parallel line means that there is *exactly* one such line.

202. Source: "Parallel line" by K. Truemper, released into Public Domain under Creative Commons CC0.

203. Source: Archiv of the Berlin-Brandenburg Academy of Sciences (ABBAW), Department Collection, Portraits of Scientists, C. F. Gauss, ZIMM-0001. Photo by Stephan Fölske. The Academy has kindly granted permission for use of the photo.
The painting has a complex history: In 1840, the Dutch painter Christian Albrecht Jensen created the Gauss portrait for the astronomical observatory in Pulkovo, a village near St. Petersburg, Russia. He also painted three copies for Johann Benedict Listing, Wilhelm Eduard Weber, and Wolfgang Sartorius von Waltershausen. Listing died 1882 in Göttingen. His copy was bought in 1883 by the National Gallery Berlin from Listing's widow and was donated in 1888 to the Prussian Academy of Science by the "Ministerium der geistlichen, Unterrichts- und Medicinalangelegenheiten." It is now owned by the Berlin-Brandenburg Academy of Sciences and Humanities.

204. Source: https://en.wikipedia.org/wiki/Carl_Friedrich _Gauss#/media/File:Carl_Friedrich_Gau%C3%9F_signature.s vg. "Carl Friedrich Gauss signature" by derivative work: Pbroks13 (talk) Carl Friedrich Gauss (1777-1855) - Carl_Friedrich_Gauss_Namenszug_von_1794.jpg. Licensed under Public Domain via Commons.

205. See Wikipedia "Constructible Polygon."

206. [Dunnington, 2004]. See Wikipedia "Carl Friedrich Gauss," including the German version.

207. For a detailed discussion of the history of area computation and the related operations, see Chapter 1 of [Dunham, 1990]. Here, we just summarize how the various operations can be carried out. Addition and subtraction are directly done with the compass. Multiplication, division, and taking of square root rely on the following drawing.

It has a baseline consisting of a and b, where in turn a is composed of r and d. The semicircle has radius r. The line segment c forms a right angle with the baseline. The dashed line segment e connects two points of the halfcircle as shown. The second dashed line goes from the midpoint of e to the center of the semicircle, and thus forms a right angle with e.

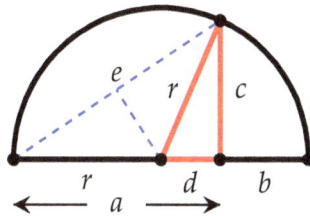

(Source: "Squared rectangle" by K. Truemper, released into Public Domain under Creative Commons CC0.)

By Pythagoras's theorem for the solid triangle with side lengths c, d, and r, we have $r^2 = c^2 + d^2$. Using $r = \frac{a+b}{2}$ and $d = a - r = \frac{a-b}{2}$, we have $r^2 = (a^2 + 2ab + b^2)/4 = c^2 + (a^2 - 2ab + b^2)/4$. Reduction of the last equation yields $c^2 = ab$.

For any two values of a, b, and c, the third value can be obtained via the above construction, as follows:

Given a and b: Without loss of generality, assume $a \geq b$. Determine $r = \frac{a+b}{2}$, draw the semicircle, draw c at right angle to base line.

Given a and c: Draw a with c at right angle, then add dashed segment e. Find the midpoint of e, erect the second dashed line at right angle, get the centerpoint of the semicircle, draw the semicircle, and thus get b. Note: The drawing is applicable only if $a \geq c$. If $a < c$, the above construction rules still apply, but a different drawing results where d is part of b instead of a.

Case of given b and c follows by symmetry with a and c.

Taking square root $\sqrt{x} = z$: Define $a = x$ and $b = 1$, then determine c. Since $x \cdot 1 = c^2$, we have $z = c$.

Multiplication $xy = z$: Use $a = x$ and $b = y$ to get c for which $c^2 = ab$. Use this c and $a = 1$ to get b for which $c^2 = 1 \cdot b$, which is the desired solution z.

Division $\frac{x}{y} = z$: Rewrite as $x = yz$, with x and y known. Determine $\sqrt{x}$. Define $a = y$ and $c = \sqrt{x}$, then determine b, which is the desired z.

208. Source: "Complex roots providing 17-sided heptadecagon" by K. Truemper, released into Public Domain under Creative Commons CC0.

209. Source: "Heptadecagon" by K. Truemper, released into Public Domain under Creative Commons CC0.

210. The only cases of interest have p odd. Otherwise, $p = 2^k \cdot q$ for some $k \geq 1$ and q odd, and a construction for the case of q immediately yields a construction for p as described earlier.

211. See Wikipedia "Complex plane" for the multiplication rules of the complex plane, which imply that the roots of $x^p - 1$ must be evenly distributed on the circle of the complex plane centered at the origin and with radius 1.

212. Gauss writes just $\frac{x^p-1}{x-1}$ when he actually means $\frac{x^p-1}{x-1} = 0$.

213. Gauss relies on the fact that $\frac{x^p-1}{x-1}$ is a polynomial whose roots are the $p - 1$ complex roots of the polynomial $x^p - 1$. This fact is a consequence of the *fundamental theorem of algebra*. See Wikipedia "Fundamental theorem of algebra" about details and the complicated history of its proofs, including four proofs by Gauss.

214. Letter to Gerling in 1819; see [Archibald, 1920].
Gauss had focused on the first complex root encountered as one traverses the unit circle counterclockwise from the single real root, which has value 1. That complex root is $R = \cos(\frac{2\pi}{17}) + \sin(\frac{2\pi}{17})i$ of the diagram. He found a way to compute that root using just the four basic arithmetic operations and taking of square root. Indeed, he established the following equation for the cosine of $\frac{2\pi}{17}$, written here in modern notation: $\cos\frac{2\pi}{17} = \frac{1}{16}[-1 + \sqrt{17} + \sqrt{34 - 2\sqrt{17}} + 2\sqrt{17 + 3\sqrt{17} - \sqrt{34 - 2\sqrt{17}} - 2\sqrt{34 + 2\sqrt{17}}}]$. Evaluation of the formula just requires repeated application of the five operations of addition, subtraction, multiplication, division, and the taking of square root. For any angle α, $\sin\alpha = \sqrt{1 - \cos^2\alpha}$, so $\sin\frac{2\pi}{17}$ can then be computed by these five operations as well. Thus, the point $R = \cos(\frac{2\pi}{17}) + \sin(\frac{2\pi}{17})i$ of the complex plane can be constructed.

215. See Wikipedia "Fermat number" for the interesting history of these primes. Chapter 10 of [Dunham, 1990] has details of Euler's proof that $2^{(2^5)} + 1 = 4,294,967,297$ is equal to $641 \cdot 6,700,417$ and thus is not a prime, at the time an astonishing achievement.

216. Source: `https://en.wikipedia.org/wiki/Pierre_de_Ferm at#/media/File:Pierre_de_Fermat.jpg`. "Pierre de Fermat" by `http://www-groups.dcs.st-and.ac.uk/~history/PictDisplay/F ermat.html`. Licensed under Public Domain via Commons.

217. See Wikipedia "Pierre Wantzel."

218. For example, the *heptagon*, with 7 edges, and the *nonagon*, with nine edges, are regular polygons that cannot be constructed. Indeed, 7 is not a Fermat prime, and 9 is equal to $3 \cdot 3$ and thus the product of two identical Fermat primes.

219. See Wikipedia "Carl Friedrich Gauss."

220. Source: `https://de.wikipedia.org/wiki/Carl_Friedrich_G au%C3%9F#/media/File:Braunschweig_Gauss-Denkmal_17-ecki ger_Stern.jpg`. "Braunschweig Gauss-Denkmal 17-eckiger Stern" by Benutzer:Brunswyk. Licensed under CC BY-SA 3.0 via Creative Commons.

221. Source: `https://en.wikipedia.org/wiki/Paolo_Ruffini# /media/File:Ruffini_paolo.jpg`. "Ruffini Paolo" by Unknown. Licensed under Public Domain via Commons.

222. Source: `https://en.wikipedia.org/wiki/Niels_Henrik_A bel#/media/File:Niels_Henrik_Abel.jpg`. "Niels Henrik Abel" by Johan Gørbitz - Originally uploaded to English wikipedia by en:User:
Pladask. Licensed under Public Domain via Commons.

223. See Wikipedia "Cubic function."

224. See Wikipedia "Quartic Function."

225. See Wikipedia "Paolo Ruffini," "Niels Henrik Abel," "Évariste Galois," and "Pierre Wantzel."

226. Source: `https://commons.wikimedia.org/wiki/File:Evar iste_galois.jpg`. Portrait by Unknown was owned by Nathalie-Théodore Chantelot, E. Galois's older sister, and her daughter Mrs. Guinard. It was released by Paul Dupuy, École Normale Supérieure

professor of history, with his article "La vie d'Évariste Galois," in 1896. "Evariste Galois" by Unknown - Iyanaga, Shokichi, Springer-Verlag Tokyo, 1999. http://www.win.tue.nl/~aeb/at/GaloisCorr espondence.html. Licensed under Public Domain via Commons.

227. A detailed account is given in [Livio, 2005].

228. See Wikipedia "Group (mathematics)."

229. The binary field is an example of a finite field. Indeed, it is the smallest field, defined as follows.

1. The numbers are 0 and 1.
2. Addition: $0 + 0 = 0, 0 + 1 = 1 + 0 = 1, 1 + 1 = 0$.
3. Subtraction: $0 - 0 = 0, 0 - 1 = 1 - 0 = 1, 1 - 1 = 0$.
4. Multiplication: $0 \cdot 1 = 1 \cdot 0 = 0, 1 \cdot 1 = 1$.
5. Division: $0/1 = 0, 1/1 = 1$.

For details about finite fields, see Wikipedia "'Finite field."

230. See Wikipedia "Finite field."

231. See Wikipedia "Pierre Wantzel" and "Angle trisection." A summary of Wantzel's arguments is included in [Cajori, 1918].

232. Assume that trisection can be done for any angle. Use trisection to divide the 360 degrees of the circle into three 120 degree slices. This was already done by the ancient Greeks via construction of the equilateral triangle.

(Source: "Trisection of circle" by K. Truemper, released into Public Domain under Creative Commons CC0.)
To each 120 degree slice, apply trisection again, getting a total of nine 40 degree slices. From these, we directly derive the nonagon with nine edges. But we have seen that this polygon cannot be constructed, so it must be impossible to carry out the second trisection step.

233. See Wikipedia "Galois theory."

234. See Wikipedia "Doubling the cube" and "Pierre Wantzel." For a summary of the arguments, see [Cajori, 1918].

235. See Wikipedia "Galois theory."

236. Source: "Triangle area" by K. Truemper, released into Public Domain under Creative Commons CC0.

237. The subdivision into triangles is carried out as follows.

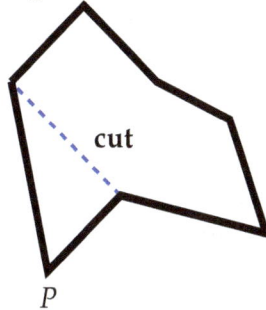

(Source: "Triangulation of polygon" by K. Truemper, released into Public Domain under Creative Commons CC0.)
Define an *ear* of the polygon to be a vertex P such that the line segment connecting the two neighbors of P lies entirely in the interior of the polygon. The two ears theorem states that a polygon has at least two ears; see Wikipedia "Two ears theorem" for details and proof. Evidently the line segment and the two edges attached to an ear P form a triangle that contains no other part of the polygon. Hence the triangle can be cut off, resulting in a polygon with one less vertex. Repeat until the polygon has become a triangle.

238. Details are provided on pp. 17–19 [Dunham, 1990]. Here is a short version.

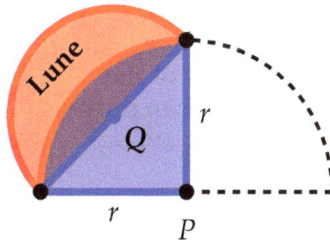

(Source: "Lune and triangle" by K. Truemper, released into Public Domain under Creative Commons CC0.)
We begin with the quarter circle that is centered at P and has radius r. It consists of a region we call the lens and a triangle where the two shorter sides form a right angle and are radii of the quarter circle. The third and longer side of the triangle is the diameter of the semicircle centered at Q and consisting of the lune and the lens.

By Pythagoras's theorem, the longer side of the triangle has length $\sqrt{r^2 + r^2} = r\sqrt{2}$. This implies that the areas of the quarter circle and the semicircle are the same; the value is $\frac{1}{4}r^2\pi$. Removing the lens they have in common, the area of the lune must be equal to that of the triangle.

239. Source: "Polygon area" by K. Truemper, released into Public Domain under Creative Commons CC0.

240. Source: "Lune of Hippocrates" by K. Truemper, released into Public Domain under Creative Commons CC0.

241. "Archimedes parabola with triangle" by K. Truemper, released into Public Domain under Creative Commons CC0.

242. See Wikipedia "Archimedes."

243. See Chapter 4.

244. [Cajori, 1918] has a summary.

245. See Chapter 2.

246. Euclid's postulates are as follows:
   1. A straight line segment can be drawn connecting any two points.
   2. Any straight line segment can be extended indefinitely in a straight line.
   3. A circle can be drawn around any point and with any radius.
   4. All right angles are equal to one another.
   5. (The parallel postulate) If a straight line intersects two straight lines such that the sum of the inner angles on one side is less than two right angles, then the two lines must intersect on that side when extended far enough.

See Wikipedia "Euclidean geometry" for further details.

247. See Wikipedia "Playfair's axiom." Note that the first four axioms imply that there is at least one parallel line. Thus, the condition of Playfair's axiom that there is *at most* one parallel line means that there is *exactly* one such line.

248. See Wikipedia "Parallel postulate."

249. Source: https://commons.wikimedia.org/wiki/File:John_Playfair_by_Sir_Henry_Raeburn.jpg. "John Playfair" by Henry

Raeburn. Public Domain under US copyright code PD-old-100.

250. An axiom *C* is implied by other axioms if *C* holds whenever the other axioms hold.

251. See Wikipedia "Parallel postulate."

252. [Engel and Stäckel, 1895] and
[Königliche Gesellschaft der Wissenschaften, Göttingen, 1900].

253. Source: `https://de.wikipedia.org/wiki/Ferdinand_Karl`
`_Schweikart#/media/File:Ferdinand_Karl_Schweikart.jpg`.
"Ferdi-
nand Karl Schweikart" by Unknown - `http://www.liveinternet.r`
`u/users/kakula/post153809122/`. Licensed under Public Domain under US copyright code PD-old-70.

254. p. 244 [Engel and Stäckel, 1895].

255. p. 246 [Engel and Stäckel, 1895].

256. p. 249, 250 [Engel and Stäckel, 1895] contains the letter of Gauss to Taurinus: "About your attempt I have little to comment except that it is incomplete. ... I guess that you have not worked for a long time on this topic. For me, it has been over 30 years, and I do not believe, that anybody has been more involved than me with the second part [concerning the sum of angles of any triangle less than 180 degrees], although I have not published anything about this. The assumption that the sum of the 3 angles is smaller than 180 degrees results in a particular, from present-day (Euclidean) different, geometry which I have worked out to my satisfaction, so that I can answer every question, except for determination of a constant which cannot be determined a priori." Gauss's letter concludes: "At any rate, you must view the above material as private information that in no way should be published or used in such a way that it could become public. Perhaps when I have more time than available at present, I will make these results public."

257. Source: `https://en.wikipedia.org/wiki/Nikolai_Lobach`
`evsky#/media/File:Lobachevsky.jpg`. "Lobachevsky" by Lev Dmitrie-
vich Kryukov. Licensed under Public Domain via Commons.

258. Source: `https://en.wikipedia.org/wiki/Henri_Poincar%C3`
`%A9#/media/File:Henri_Poincar%C3%A9-2.jpg`. "Henri Poincaré-
2" by Unknown - `http://www.mlahanas.de/Physics/Bios/image`

s/HenriPoincare.jpg. Licensed under Public Domain via Commons.

259. Source: "Poincaré disk" by K. Truemper, released into Public Domain under Creative Commons CC0.

260. Indeed, it is easy to see that there are an infinite number of lines going through point P that are parallel to line A. Thus, Lobachevsky's condition that there are at least two parallel lines implies that, for any given line and any point not on that line, there are an infinite number of parallel lines going through that point.

261. Source: http://www.titoktan.hu/Bolyai_a.htm. "János Bolyai" photo by copyright holder Támas Dénes, who has kindly granted permission to use the photo. According to [Dénes, 2011] and the Wikipedia "János Bolyai" entry, no original portrait of Bolyai survives, and an unauthentic picture appears in some encyclopedias and on a Hungarian postage stamp.
The relief of János Bolyai shown here is part of six reliefs in front of the Culture Palace in Marosvásárhely, Romania. According to the investigation reported in [Dénes, 2011] that included computer simulation using images across three generations, the relief likely is a good, indeed the only authentic, representation of the mathematician.

262. See Wikipedia "Non-Euclidean geometry."

263. See Wikipedia "János Bolyai."

264. Gauss's evaluation is confirmed by recent investigations into the work of János Bolyai; see [Dénes, 2011] and [Kiss, 1999].

265. Gauss's letter to Taurinus includes the following philosophical statement, see p. 250 [Engel and Stäckel, 1895]:
". . . All efforts of mine to find a contradiction, an inconsistency in this non-Euclidean geometry have been fruitless, and the only thing that makes our brain resist it [the geometry], is that, if true, there would have to be a certain (though not known by us) constant of linearity. But I think that, despite the vacuous word-wisdom of the metaphysicists, we know not enough or indeed nothing about the true nature of space, that we can permit us to confuse something that appears to us unnatural with something that is absolutely impossible. If the Euclidean geometry was the true one, and the constant [used in the non-Euclidean geometry] was related to values that could be measured on earth or up in the sky, then we could determine the true nature [of space] *a posteriori*. So jokingly I have

sometimes expressed the wish that the Euclidean geometry was not the true one, since then we would have an absolute measure [of the constant] a priori."

266. Source: http://www-groups.dcs.st-and.ac.uk/history/ PictDisplay/Beltrami.html. "Eugenio Beltrami" by Unknown. Public Domain under US copyright code PD-old-70.

267. See Wikipedia "Eugenio Beltrami."

268. See Wikipedia "Elliptic geometry."

269. Source: https://en.wikipedia.org/wiki/Elliptic_geometr y#/media/File:Triangles_%28spherical_geometry%29.jpg. "Triangles (spherical geometry)" by Lars H. Rohwedder, Sarregouset - Own work from source files Image:OgaPeninsulaAkiJpLandsat.jpg (GFDL) and Image:Orthographic Projection Japan.jpg (GFDL and CC-By-SA). Licensed under CC BY-SA 3.0 via Commons.

270. See Wikipedia "Great circle."

(Source: https://en.wikipedia.org/wiki/Great_circle#/medi a/File:Great_circle_hemispheres.png. "Great Circle" by Jhbdel, en.wikipedia. Licensed under CC BY-SA 3.0 via Commons.)

271. Technically, any two antipodal points are *identified* to represent one point of the geometry. This is needed so that Euclid's axiom is satisfied that two points uniquely determine a line. See Wikipedia "Elliptic geometry."

Chapter 6 Proof

272. For example, the website snopes.com for "urban legends, folklore, myths, rumors, and misinformation" can be helpful.

273. See Wikipedia "Occam's razor."

274. Source: https://en.wikipedia.org/wiki/Albert_Einstein# /media/File:Albert_Einstein_%28Nobel%29.png. "Albert Einstein (Nobel)" by Unknown - Official 1921 Nobel Prize in Physics pho-

tograph. Licensed under Public Domain via Commons.

275. Source: https://en.wikipedia.org/wiki/Minkowski_space# /media/File:GPB_circling_earth.jpg. "GPB circling earth" by NASA - http://www.nasa.gov/mission_pages/gpb/gpb_012.html. Licensed under Public Domain via Commons.

276. See Wikipedia "Space-time."

277. Source: https://www.math.ubc.ca/~cass/Euclid/ybc/ybc .html. Permission for use granted by William A. Casselman, photographer and copyright holder. The clay tablet is part of the Yale Babylonian Collection. Provenance unknown, dated approximately 1800–1600 BCE. Purchased around 1912 by an agent of J. P. Morgan, who contributed it to Yale University as part of the foundation of its Babylonian Collection.

278. Source: https://en.wikipedia.org/wiki/File:Plimpton_3 22.jpg. "Plimpton 322" by photo author unknown. Licensed under Public Domain via Commons.

279. See Wikipedia "Pythagorean theorem" for various proofs.

280. p. 215 [Rudman, 2007].

281. [Ossendrijver, 2016].

282. p. 21 [Wilson, 1828].

283. See Wikipedia "Apparent retrograde motion."

284. Source: "Jupiter velocity graph" by K. Truemper, released into Public Domain under Creative Commons CC0. The drawing is based on a figure of [Ossendrijver, 2016].

285. Source: Photo by Mathieu Ossendrijver, Humboldt University/ British Museum. M. Ossendrijver has kindly granted permission for use of the photo.

286. [Ossendrijver, 2016].

287. See Wikipedia "Aristotle."

288. See Wikipedia "Euclid."

289. Source: https://commons.wikimedia.org/wiki/File:Aristo tle_Altemps_Inv8575.jpg. "Aristotle Altemps Inv8575" by Copy of Lysippus - Jastrow (2006). Licensed under Public Domain via

Commons.

290. Here are some of the accomplishments of Archimedes; for details see [Netz and Noel, 2007]:

1. An iterative scheme to compute π with as high a precision as desired, using approximating polygons; he carried out that method for the 96-sided polygon, getting the bounds $3\frac{10}{71} < \pi < 3\frac{1}{7}$, with midpoint estimate 3.14185. The estimate matches three digits after the decimal point of the actual value $\pi = 3.14159\ldots$.

2. Computation of areas and volumes with a precursor of integral calculus, see Chapters 3 and 5.

3. Schematics for geometric problem that do not correspond to reality but show logic relationships of points and lines.

4. Counting of geometric configurations.

291. For the period up to about 100 BCE, any list should include Thales of Miletus, Pythagoras of Samos, Zeno of Elea, Hippocrates of Chios, Archytas of Tarentum, Theaetetus of Athens, Eudoxus of Cnidus, Apollonius of Perga, and Hipparchus of Nicaea. Wikipedia has detailed information for each mathematician of the list.

292. [Lenzen, 2004].

293. Leibniz anticipated results for the following areas of modern logic.

1. *Propositional logic*: It uses *propositional variables* with values *True* and *False* to encode elementary facts and expresses their relationships with the basic *operators* "not," "and," "or," plus advanced operators such as "if ... then" and "only if." For example, let the variable x have value *True* if it is raining, and define variable y to have that value if a person uses an umbrella. Then the formula *if x then y* has value *True* if it is not raining or the person uses an umbrella.

2. *First-order logic*: It uses *predicates*, also called *truth functions*, in addition to propositional variables. An example is the function $woman(x)$ where x represents an arbitrary person of the human population and $woman(x)$ has value *True* if x is a woman and *False* otherwise. For the description of relationships among predicates, the operators of propositional logic are used plus the *quantifiers* "exists" and "for all." For example, the compact statement *exists x woman(x)* says that there is a person on

earth who is a woman. For details, see Wikipedia "First-order
logic."

3. *Modal logic*: It generalizes first-order logic by allowing operators such as "possibly," "necessarily," "impossibly," "it is
obligatory that," and "it is permissible that." For example,
the statement "The woman possibly was an airline employee"
can be expressed in modal logic. For details, see Wikipedia
"Modal logic."

294. [Peckhaus, 1997].

295. See Wikipedia "Calculus ratiocinator."

296. See Wikipedia "Stepped reckoner."

297. See Wikipedia "Characteristica universalis."

298. See Wikipedia "Artificial intelligence."

299. See Wikipedia "Calculus ratiocinator" and "Characteristica universalis."

300. See [Morgan, 1847] and p. 1863, 1864 [Newman, 1956].

301. Source: https://en.wikipedia.org/wiki/File:De_Morgan_
Augustus.jpg. "Memoir of Augustus de Morgan" by Sophia Eliz-
abeth De Morgan, 1882. Licensed under Public Domain via Commons.

302. See pp. 1864–1868 [Newman, 1956]. Boole's approach may be
sketched as follows.
He starts with classes having certain features. Let x, y, z denote
such classes. The class containing all features, the *universe of discourse*, is denoted by 1, and that class not containing any feature,
the *null class*, is denoted by 0. The class having the features of both
x and y is xy. For x and y with disjoint features, $x + y$ is the class
where each element has either the features of x or those of y. Finally, if $z = x + y$, then $x = z - y$ is declared to hold.
Since 1 is the universe of discourse and 0 is the null class, $1x = x$
and $0x = 0$. The complement of x is $1 - x$. These definitions cause
problems: For example, $x + x$ makes sense only if x is the empty
class, and $x - y$ is defined only if x contains y. But then Boole defines a specialized version by adding the condition that each class
must be equal to 0 or 1. Boole handles the undefined $x + x$ by allowing integers larger than 1 during computations. The latter change
has the drawback that intermediate steps of calculations may not

have an interpretation in terms of classes. Nevertheless, Boole's system makes reliable logic calculations possible.

303. Source: `https://en.wikipedia.org/wiki/George_Boole#/media/File:George_Boole_color.jpg`. "George Boole color" by Unknown. Licensed under Public Domain via Commons.

304. Source: `https://archive.org/`. Search for "Boole Laws of Thought." Licensed under Public Domain Mark 1.0 via Creative Commons.

305. p. 69, 70 [Boole, 1854].

306. The exclusive "or" was replaced by the inclusive version, and the subtraction operator was dropped.

307. See Wikipedia "Boolean algebra."

308. See Wikipedia "Boolean algebra."

309. [Frege, 1879].

310. Besides propositional logic and first-order logic cited earlier, Frege also defined *second-order logic* where quantifiers "exists" and "for all" may not only refer to a universe of discourse, but may apply to predicate functions. One may re-express this by saying that second-order logic allows quantification over sets instead of just individual members of the universe. An example statement is *for every subset P of the universe and every member x of the universe, x is in P or x is not in P.*

311. Source: `https://en.wikipedia.org/wiki/Gottlob_Frege#/media/File:Young_frege.jpg`. "Young Frege" by Unknown. Licensed under Public Domain via Commons.

312. Source: `https://en.wikipedia.org/wiki/Begriffsschrift#/media/File:Begriffsschrift_Titel.png`. Uses digitized version at `http://gallica.bnf.fr/ark:/12148/bpt6k65658c`. Licensed under Public Domain via Commons.

313. [Frege, 1893].

314. [Frege, 1884].

315. p. 1 [Frege, 1893].

316. Source: `https://archive.org/`. Search for "Gottlob Frege Grundgesetze der Arithmetik." Public Domain.

317. The appendix opens with the statement, "To a science writer there is hardly anything as undesirable as the realization at the end of a project that one of the foundations of the building has been shattered." It concludes with, "As fundamental problem of arithmetic one can view to be the question: How do we capture the logic elements, in particular the numbers? Why are we justified to interpret the numbers as things? Even though that problem is not solved to the extent as I had thought during the writing of this volume, I still have no doubt that the road to success has been identified."

318. Source: `https://en.wikipedia.org/wiki/Bertrand_Russe ll#/media/File:Russell_in_1938.jpg`. "Bertrand Russell" by Unknown. Licensed under Public Domain via Commons.

319. The following discussion of the contradiction inherent in Frege's set definition is based on Wikipedia "Russell's paradox." Define a set to be *normal* if it does not contain itself as an element, and to be *abnormal* otherwise. For example, the set of spoons is itself not a spoon, and thus is normal. On the other hand, the set where each element is not a spoon, itself isn't a spoon either, and thus is abnormal.
Now define S to be the set of all normal sets. Is S normal or abnormal? Suppose S is normal and thus does not contain itself as an element. But that contradicts that S contains all normal sets. Now suppose S is abnormal and thus contains itself as an element. But by definition of S, all elements of S are normal. Thus S is normal, another contradiction.
Since both cases result in a contradiction, the set S cannot exist. But Frege's construction of sets via logic conditions declares S to exist, which can only mean that the construction is inherently faulty.

320. See Wikipedia "David Hilbert."

321. See Wikipedia "Hilbert's problems" for details, including current status of solved versus still unsolved.

322. Source: `https://en.wikipedia.org/wiki/David_Hilbert#/m edia/File:Hilbert.jpg`. "David Hilbert" by Unknown. Licensed under Public Domain via Commons.

323. The definition of well-ordering is based on a feature of the set N of natural numbers: Take any subset S of N containing at least one number; then among the numbers of S, one number is smallest. Based on that feature of N, a general set with a given

ordering is said to be *well ordered* if any nonempty subset has a smallest element.

Now some sets are not well ordered, for example, the set of integers with their natural ordering. Indeed, the subset consisting of -1, -2, $-3\ldots$ has no smallest element. But the integers *can* become well ordered when we use a different ordering. For example, we can declare the integers to have the increasing order 0, -1, 1, -2, 2, -3, $3\ldots$, where we are suspending the usual interpretation of the numbers. Then every subset does have a smallest element.

324. Source: https://en.wikipedia.org/wiki/Ernst_Zermelo#/media/File:Ernst_Zermelo.jpeg. "Ernst Zermelo" by Konrad Jacobs - http://owpdb.mfo.de/detail?photo_id=8666. Licensed under CC BY-SA 2.0 de via Commons.

325. Source: https://en.wikipedia.org/wiki/Basket#/media/File:Gullah_basket.JPG. "Gullah basket" by Bubba73 (Jud Mc-Cranie) - Own work. Licensed under CC BY-SA 4.0 via Commons.

326. See Wikipedia "Axiom of countable choice."

327. Source: https://en.wikipedia.org/wiki/Abraham_Fraenkel#/media/File:Adolf_Abraham_Halevi_Fraenkel.jpg. "Adolf Abraham Halevi Fraenkel" by The David B. Keidan Collection of Digital Images from the Central Zionist Archives (via Harvard University Library). Licensed under Public Domain via Commons.

328. See Wikipedia "Banach-Tarski paradox."

329. Source: https://en.wikipedia.org/wiki/Banach%E2%80%93Tarski_paradox#/media/File:Banach-Tarski_Paradox.svg. "Banach-Tarski Paradox" by Benjamin D. Esham. Licensed under Public Domain via Commons.

330. Source: Institute of Mathematics of the Polish Academy of Sciences, which kindly granted use of the photo.

331. Source: http://owpdb.mfo.de/detail?photo_id=6091. "AlfredTarski1968" by George M. Bergman - The Oberwolfach photo collection. Licensed under Commons CC BY-SA 2.0 DE.

332. See Wikipedia "J. E. L. Brouwer."

333. Source: "BrKop7_3.jpg" photo of Brouwer Archive, National Archive Haarlem. The Archive kindly has granted use of the photo.

334. See Wikipedia "Topology."

335. The theorem says that, for any compact convex set S in any n-dimensional Euclidean space and any continuous function f that maps S into itself, there is a point $x \in S$ such that $f(x) = x$. The point x is called a *fixed point* of f. For details and other fixed-point theorems, see Wikipedia "Brouwer's fixed-point theorem" and "Fixed-point theorem."

336. See Wikipedia "Fixed-point theorem."

337. See Wikipedia "Intuitionism."

338. Aristotle's law of the excluded middle states that, for any proposition, either the proposition is true or it is false. Thus, there is no third case possible.

339. Here are two additional examples based on discussion on p. 47, 48 [Weyl, 1921], reprinted p. 165, 166 [Thiel, 1982]:
Suppose we have established an infinite collection of real numbers x, all of which are less than some given number, say 1. The numbers x may not be directly given, but indirectly specified by some process or theorem. The axiomatic method then allows the claim that there is a real number, say y, with the following property: All numbers x are less than or equal to y, and there is no number z smaller than y for which this can be claimed. The number y is called the *least upper bound* for the x numbers.
The intuitionist will not accept the claim or the use of such y unless a constructive procedure is offered. In particular, the Dedekind cut of Chapter 2, which invokes such bounds y en masse while creating the real numbers, is rejected outright by the intuitionist.
As second example, the intuitionist rejects the claim that every infinite set contains a countable infinite subset. Since that result is a consequence of the axiom of countable choice—see Wikipedia "Axiom of countable choice"— that axiom as well as the general axiom of choice are rejected as well.

340. [Weyl, 1921], reprinted pp. 157–178 [Thiel, 1982]. See also Wikipedia "Hermann Weyl."

341. Source: https://en.wikipedia.org/wiki/Hermann_Weyl#/media/File:Hermann_Weyl_ETH-Bib_Portr_00890.jpg. "Hermann Weyl" by ETH Zürich - ETH-Bibliothek Zürich, Bildarchiv. Licensed under CC BY-SA 3.0 via Commons.

342. For details of the struggle, see Stanford Encyclopedia of Philosophy "Luitzen Egbertus Jan Brouwer."

343. When *ZF* is assumed, the axiom of choice can be used to prove the well-ordering theorem, and vice versa. As a result, *ZFC* is sometimes stated with the well-ordering theorem instead of the axiom of choice. For example, see Wikipedia "Zermelo-Fraenkel set theory."

344. See Wikipedia "L. E. J. Brouwer."

345. See Wikipedia "Hermann Weyl."

346. The entire section is based on Wikipedia "Principia Mathematica."

347. Source: HUP Whitehead, Alfred North (3b), W395042_1, Harvard University Archives. "Alfred North Whitehead" photo is part of collection of photos of Harvard faculty and buildings taken in 1936 by Richard Carver Wood. Harvard University kindly granted permission to use the photo.

348. Source: `https://archive.org/`. Search for "Principia Mathematica Whitehead Russell." Public Domain.

349. For example, the very first statement defines the implication *p implies q* by the statement *not p or q*.

350. See Wikipedia "Principia Mathematica."

351. In terms of logic: A systems of axioms is consistent if there is no statement *S* such both *S* and the negation of *S* can be proved.

352. Generally, completeness and consistency are difficult to establish. For example, Gauss hesitated to publish his results on hyperbolic geometry partly because he could not establish consistency; see Chapter 5. Beltrami then proved that hyperbolic geometry is consistent if and only if this is true for Euclidean geometry. This relative result was replaced by absolute statements about consistency of Euclid's geometry, and thus of hyperbolic geometry, in the 20th century; see Wikipedia "Euclidean geometry."

353. Source: `https://en.wikipedia.org/wiki/Emil_du_Bois-Rey mond#/media/File:Emil_DuBois-Reymond_BNF_Gallica_crop.jpg`. "Emil du Bois-Reymond" by Haase phot. Berlin. Upload, stitch and restoration by Jebulon - Bibliothèque Nationale de France. Public Domain.

354. See Wikipedia "Ignoramus et ignorabimus."

355. See Wikipedia "Ignoramus et ignorabimus." Hilbert read on

the German radio an abbreviated four-minute version of the speech. The recording is available at http://www.maa.org/press/periodic als/convergence/david-hilberts-radio-address, together with the German text as well as an English translation. The speech concludes with the cited passage.

356. Source: https://en.wikipedia.org/wiki/David_Hilbert#/m edia/File:G%C3%B6ttingen_Stadtfriedhof_Grab_David_Hilbert. jpg. "Hilbert Grab" by Kassandro - Own work, https://commons. wikimedia.org/w/index.php?curid=4219496. Licensed under CC BY-SA 3.0 via Commons.

357. See Wikipedia "Hilbert's program" for details.

358. We cannot include even a summarizing discussion of the rules of finitary proofs as proposed by Hilbert and later amended by others. The principal idea is that the proof steps cannot invoke infinite processes. But formalizing this idea is another matter, and to date no complete set of rules for finitary proofs has been stated. For details about the various viewpoints, see Stanford Encyclopedia of Philosophy "Hilbert's program."

359. See Stanford Encyclopedia of Philosophy "Hilbert's program."

360. Source: https://en.wikipedia.org/wiki/Kurt_G%C3%B6del# /media/File:Kurt_g%C3%B6del.jpg. "Kurt Goedel, ca. 1926" by Unknown. Licensed under Public Domain via Commons.

361. For references and details, see Stanford Encyclopedia of Philosophy "Gödel's Incompleteness Theorems."

362. This condition is not exactly the same as in the first incompleteness theorem. Indeed, it is a bit stronger than in the first incompleteness theorem, where the condition is very weak.

363. It may appear that the condition "contains a certain amount of elementary arithmetic" may be severe enough that a number of mathematical systems are not affected by the two incompleteness theorems. But that condition can be weakened as long as the system contains parts that can be *interpreted* as a certain amount of elementary arithmetic.

364. See Chapter 4.

365. Source: https://de.wikipedia.org/wiki/Datei:Max_Plan ck_%281858-1947%29.jpg. "Max Planck" by Unknown - http: //www.sil.si.edu/digitalcollections/hst/scientific-ide

ntity/CF/display_results.cfm?alpha_sort=p. Licensed under Public Domain via Commons.

366. One can express the two conditions for independence in the terminology of theorems as follows. Requiring consistency of system S with axiom A added is equivalent to demanding that $A =$ false cannot be proved to be a theorem of S. Calling for consistency of S with the negation of A added is the same as demanding that $A =$ true cannot be proved to be a theorem of S.

367. See Stanford Encyclopedia of Philosophy "Kurt Gödel."

368. See Wikipedia "Paul Cohen."

369. Source: Courtesy Cohen Family/Stanford University, who kindly granted permission to use the photo.

370. See Wikipedia "Paul Cohen." It may seem odd that Gödel compared the effect the proof had on him with his reaction to a really good play. But brain science has shown that the area of the brain evaluating the beauty of mathematical statements is also active when responding to visual, musical, and even moral beauty; see [Zeki et al., 2014].

371. See Wikipedia "Forcing (mathematics)."

372. [Wolchover, 2013].

373. See Wikipedia "Nicolas Bourbaki." Professor Bourbaki supposedly works at the University of Nancago, a name composed from *Nancy*, France, and *Chicago*, USA. One is tempted to guess that Nancago is situated halfway between these two cities and thus in the middle of the Atlantic Ocean. Maybe it is the Atlantis of lore?

374. [Bourbaki, 1948].

375. See Wikipedia "Fermat's last theorem."

376. See Wikipedia "Fermat's last theorem."

377. See Wikipedia "Fermat's last theorem" for details of the historical developments.

378. See Wikipedia "Andrew Wiles."

379. See Wikipedia "Modularity theorem."

380. Source: https://en.wikipedia.org/wiki/Andrew_Wiles#/me

dia/File:Andrew_wiles1-3.jpg. "Andrew Wiles" copyright C. J. Mozzochi, Princeton N.J - http://www.mozzochi.org/deligne60/Deligne1/_DSC0024.jpg. The copyright statement grants free use.

381. See Wikipedia "Modularity theorem."

382. [Singh, 1997].

383. For example, the Langlands Program seeks to establish profound connections between number theory and geometry. See Wikipedia "Langlands program."

Bibliography

[Alexander, 2014] Alexander, A. (2014). *Infinitesimal: How a Dangerous Mathematical Theory Shaped the Modern World*. Scientific American/Farrar, Straus, and Giroux.

[Archibald, 1920] Archibald, R. C. (1920). Gauss and the Regular Polygon of Seventeen Sides. *American Mathematical Monthly*, vol. 27.

[Boole, 1854] Boole, G. (1854). *An Investigation of the Laws of Thought on which are founded the Mathematical Theories of Logic and Probabilities*. Walton & Maberly; go to `https://archive.org/index.php` and search for "George Boole Laws of Thought".

[Bos, 1974] Bos, H. J. M. (1974). Differentials, higher-order differentials and the derivative in the Leibnizian calculus. *Archive for History of Exact Sciences*, vol. 14.

[Bourbaki, 1948] Bourbaki, N. (1948). *L'Architecture des Mathématiques*. In *Les grands courants de la pensée mathématique*, Cahiers du Sud, Marseille.

[Bürgi, 1620] Bürgi, J. (1620). *Aritmetische und Geometrische Progress Tabulen*. Paul Sessen, Universitätsbuchdruckerei, Prag; go to `http://daten.digitale-sammlungen.de/~db/0008/bsb00082065/images/index.html?id=00082065&groesser=&fip=193.174.98.30&no=&seite=7`.

[Cajori, 1918] Cajori, F. (1918). Pierre Laurent Wantzel. *Bull. Amer. Math. Soc.*, vol. 24.

[Cajori, 1919a] Cajori, F. (1919a). *A History of Mathematics*. Second edition, revised and enlarged. Macmillan Company; go to `ht tps://archive.org/index.php` and search for "A History of Mathematics Florian Cajori".

[Cajori, 1919b] Cajori, F. (1919b). *A History of the Conceptions of Limits and Fluxions in Great Britain*. Open Court Publishing Company; go to `https://archive.org/index.php` and search for "Limits and Fluxions Florian Cajori".

[Cajori, 1928] Cajori, F. (1928). *A History of Mathematical Notations, vol. I: Notations in Elementary Mathematics*. Open Court Publishing Company; go to `https://archive.org/details/in.ernet .dli.2015.200372/mode/2up`.

[Dedekind, 1872] Dedekind, R. (1872). *Stetigkeit und irrationale Zahlen*. Friedrich Bieweg und Sohn; go to `https://archive.or g/index.php` and search for "Stetigkeit und irrationale Zahlen".

[Dénes, 2011] Dénes, T. (2011). Real Face of János Bolyai. *Notices of the American Mathematical Society*, Jan. 2011.

[Descartes, 1637] Descartes, R. (1637). *Discours de la méthode (Discourse on the Method)*. Ian Maire. Go to `http://www.gutenberg. org/` and search for "Discourse of the Method" for the English version.

[Dreyfus, 1965] Dreyfus, S. E. (1965). *Dynamic Programming and the Calculus of Variations*. Academic Press.

[Dunham, 1990] Dunham, W. (1990). *Journey through Genius: The Great Theorems of Mathematics*. Wiley.

[Dunnington, 2004] Dunnington, G. W. (2004). *Carl Friedrich Gauss: Titan of Science*. Mathematical Association of America.

[Engel and Stäckel, 1895] Engel, F. and Stäckel, P. (1895). *Die Theorie der Parallellinien von Euklid bis auf Gauß*. Teubner Verlag; available at `https://archive.org/details/theoriederparall 00stac`.

[Erdös and Dudley, 1983] Erdös, P. and Dudley, U. (1983). Some remarks and problems in number theory related to the work of Euler. *Mathematics Magazine*, vol. 56.

[Frege, 1879] Frege, G. (1879). *Begriffsschrift, eine der arithmetischen nachgebildete Formelsprache des reinen Denkens*. Verlag von Louis Nebert. English translation available at `http://dec59.ruk.cuni.cz/~kolmanv/Begriffsschrift.pdf`.

[Frege, 1884] Frege, G. (1884). *Die Grundlagen der Arithmetik: Eine logisch mathematische Untersuchung über den Begriff der Zahl*. Verlag von Wilhelm Koebner; go to `https://archive.org/index.php` and search for "Gottlob Frege Grundlagen der Arithmetik".

[Frege, 1893] Frege, G. (1893). *Grundgesetze der Arithmetik, Begriffsschriftlich abgeleitet*. Verlag von Hermann Pohle; available at `http://gallica.bnf.fr/ark:/12148/bpt6k77790t/f3.image`.

[Fritz, 1945] Fritz, K. v. (1945). The discovery of incommensurability by Hippasus of Metapontum. *Annals of Mathematics*, vol. 46.

[Gieswald, 1856] Gieswald, H. R. (1856). *Justus Byrg als Mathematiker und dessen Einleitung in seine Logarithmen*. St. Johannisschule, Danzig, Prussia; available at Bayerische Staatsbibliothek `http://mdz-nbn-resolving.de/urn:nbn:de:bvb:12-bsb10979407-8`.

[Grötschel et al., 2016] Grötschel, M., Knobloch, E., Schiffers, J., Woisnitza, M., and Ziegler, G. M., editors (2016). *Leibniz: Vision als Aufgabe*. Berlin-Brandenburgische Akademie der Wissenschaften.

[Heath, 1910] Heath, T. L. (1910). *Diophantus of Alexandria: A Study in the History of Greek Algebra*. Cambridge University Press; go to `https://archive.org/index.php` and search for "Heath Diophantus of Alexandria".

[Kiss, 1999] Kiss, E. (1999). *Mathematical Gems from the Bolyai Chests*. Akadémiai Kiadó, Budapest; Typotex LTD, Budapest.

[Königliche Gesellschaft der Wissenschaften, Göttingen, 1900] Königliche Gesellschaft der Wissenschaften, Göttingen (1900).

Carl Friedrich Gauss Werke. Teubner Verlag; available as ebook at `http://gdz.sub.uni-goettingen.de/dms/load/img/?PPN=PPN2 36010751&IDDOC=136917`.

[Lenzen, 2004] Lenzen, W. (2004). *Calculus Universalis: Studien zur Logic von G. W. Leibniz.* Mentis Verlag.

[Livio, 2005] Livio, M. (2005). *The Equation That Couldn't Be Solved.* Simon & Schuster.

[Morgan, 1847] Morgan, A. D. (1847). *Formal Logic: or The Calculus of Inference, Necessary and Probably.* Taylor and Walton; go to `https://archive.org/index.php` and search for "De Morgan Formal Logic".

[Netz and Noel, 2007] Netz, R. and Noel, W. (2007). *The Archimedes Codex.* Weidenfeld & Nicolson, paperback Phoenix.

[Newman, 1956] Newman, J. R. (1956). *The World of Mathematics,* Vols. I-IV. Simon & Schuster; go to `https://archive.org/inde x.php` and search for "james newman world of mathematics".

[Ossendrijver, 2016] Ossendrijver, M. (2016). Ancient Babylonian astronomers calculated Jupiter's position from the area under a time-velocity graph. *Science*, vol. 351.

[Peckhaus, 1997] Peckhaus, V. (1997). *Mathesis universalis und allgemeine Wissenschaft. Leibniz und die Wiederentdeckung der formalen Logik im 19. Jahrhundert.* Akademie-Verlag (Logica Nova).

[Rudman, 2007] Rudman, P. S. (2007). *How Mathematics Happened.* Prometheus Books.

[Sautoy, 2003] Sautoy, M. d. (2003). *The Music of the Primes: Searching to Solve the Greatest Mystery in Mathematics.* HarperCollins.

[Schoenflies, 1927] Schoenflies, A. (1927). Die Krisis in Cantor's mathematischem Schaffen. *Acta Mathematica*, vol. 50.

[Singh, 1997] Singh, S. (1997). *Fermat's Enigma: The Epic Quest to Solve the World's Greatest Mathematical Problem.* Walker.

[Stifel, 1544] Stifel, M. (1544). *Arithmetica Integra.* Johannes Petreius, Nürnberg.

[Thiel, 1982] Thiel, C. (1982). *Erkenntnistheoretische Grundlagen der Mathematik*. Gerstenberg Verlag.

[Truemper, 2017] Truemper, K. (2017). *The Construction of Mathematics: The Human Mind's Greatest Achievement*. Leibniz Company.

[Truemper, 2020] Truemper, K. (2020). *The Daring Invention of Logarithm Tables: How Jost Bürgi, John Napier, and Henry Briggs Simplified Arithmetic and Started the Computing Revolution*. Leibniz Company.

[Truemper, 2022] Truemper, K. (2022). *Wittgenstein and Brain Science: Understanding the World*. Leibniz Company.

[Truemper, 2024] Truemper, K. (2024). *Die wagemutige Erfindung der Logarithmentafeln: Wie Jost B"urgi, John Napier und Henry Briggs das Rechnen revolutionierten*. Leibniz Company.

[Waldvogel, 2014] Waldvogel, J. (2014). Jost Bürgi and the discovery of the logarithms. *Elemente der Mathematik*, vol. 69, pp. 89–117.

[Weyl, 1921] Weyl, H. (1921). Über die neue Grundlagenkrise der Mathematik. *Mathematische Zeitschrift*, vol. 10.

[Wilson, 1828] Wilson, J. (1828). *The Tetrabiblos; or, Quadripartite of Ptolemy*. William Hughes; see Google books "wilson the tetrabiblos".

[Wolchover, 2013] Wolchover, N. (2013). Dispute over Infinity Divides Mathematicians. *Quanta Magazine*, December, 2013.

[Zeki et al., 2014] Zeki, S., Romaya, J. P., Benincasa, D. M. T., and Atiyah, M. F. (2014). The experience of mathematical beauty and its neural correlates. *Frontiers in Human Neuroscience*, vol. 8.

Acknowledgements

This book is based on my previous book *The Construction of Mathematics* ([Truemper, 2017]), which investigates whether mathematics is created or discovered. Here we ignore that question and solely focus on the struggle of mathematicians to achieve clarity and consistency in mathematics in such a way that the reader acquires a fundamental understanding of key concepts of mathematics.

The following persons and institutions provided material and results: Academy of Sciences Berlin-Brandenburg, I. Adler, Biblioteka IM PAN Warsaw, W. A. Casselman, Cohen family, A. Czader, D. van Dalen, T. Dénes, V. Enke, D. K. Frasier, I. Grötschel, M. Grötschel, Harvard University, Institute for Advanced Study Princeton, S. Janeczko, H. Khelif, E. Knobloch, J. Koblitz, W. Lenzen, Lilly Library of University of Indiana, M. Ossendrijver, V. Peckhaus, Polish Academy of Sciences Warsaw, RAND Corporation, T. Reggev, F. Staudacher, Stanford University, University of Texas at Dallas, and J. Waldvogel.

Ingrid and Ute Truemper were patient editors.

We very much thank all these persons and institutions for their help.

K. T.

Index

www.ingramcontent.com/pod-product-compliance
Lightning Source LLC
Chambersburg PA
CBHW050116210326
41519CB00015BA/3984